Books by Richard Perry

THE WORLD OF THE TIGER
THE WORLD OF THE POLAR BEAR
THE WORLD OF THE WALRUS
THE WORLD OF THE GIANT PANDA
THE WORLD OF THE JAGUAR
AT THE TURN OF THE TIDE

The Many Worlds of Wildlife Series

THE UNKNOWN OCEAN
THE POLAR WORLDS
LIFE AT THE SEA'S FRONTIERS

LIFE AT THE SEA'S FRONTIERS

LIFE AT THE

Illustrated by Nancy Lou Gahan

SEA'S FRONTIERS

Richard Perry

VOLUME III : THE MANY WORLDS OF WILDLIFE SERIES

Taplinger Publishing Company   *New York*

First Edition
Published in the United States in 1974 by
TAPLINGER PUBLISHING CO., INC.
New York, New York

Lithographed in the United States by
Vail-Ballou Press, Inc., Binghamton, New York

Published simultaneously in the Dominion of Canada by
Burns & MacEachern, Ontario

Library of Congress Catalog Card Number: 73-3969

ISBN 0-8008-4795-4

Designed by Mollie M. Torras

Galápagos: Island of Birds by Bryan Nelson. Copyright © 1968. Reprinted with permission of Longman Group Limited.

Galápagos: World's End by William Beebe. Copyright 1924 by G. P. Putnam's Sons; copyright renewed 1951 by William Beebe. Reprinted with the permission of G. P. Putnam's Sons.

Inagua: Which Is the Name of a Very Lonely and Nearly Forgotten Island by Gilbert Klingel. Copyright 1940 by Dodd, Mead & Company, Inc.; copyright renewed 1968 by Gilbert Klingel. Reprinted with the permission of Dodd, Mead & Company, Inc.

Journey to Red Birds by Jan Linblad. Copyright © 1969. Reprinted with the permission of Farrar, Straus & Giroux, Inc.

Life and Death in a Coral Sea by Jacques-Yves Cousteau and Philippe Diolé. Copyright © 1971 by Jacques-Yves Cousteau. Reprinted with the permission of Doubleday & Company, Inc.

The Reptiles by Archie Carr and the Editors of Time-Life Books. Copyright © 1963 by Time, Inc. Reprinted with the permission of Time-Life Books.

Seals by K. M. Backhouse. Copyright © 1969. Reprinted with the permission of Arthur Barker Ltd.

The Shoals of Capricorn by Francis D. Ommanney. Copyright 1952. Reprinted with the permission of Longman Group Limited.

Song of Wild Laughter by Jack Couffer. Copyright © 1955 by Jack Couffer. Reprinted with the permission of Simon & Schuster, Inc.

Transactions of the Natural History Society of Northumberland, Durham & Newcastle upon Tyne 11, no. 9 (1956). The extract by Eric Simms is reprinted with the permission of the Natural History Society of Northumberland, Durham and Newcastle upon Tyne.

Zoo Quest for a Dragon by David Attenborough.. Copyright © 1957. Reprinted with the permission of Lutterworth Press.

for Sam

ACKNOWLEDGEMENTS

To Professor Walter Auffenberg of the Florida State Museum, Dr F. Wayne King of the New York Zoological Society and Professor Paul A. Colinvaux of Ohio State University, for being so kind as to answer my queries and also to send me papers; to the Secretary of the Natural History Society of Northumberland, Durham & Newcastle upon Tyne, for permission to quote from the Society's *Transactions*, and to the Editor of the U.S. National Museum's *Bulletin* for permission to quote from *Life Histories of North American Shore Birds*; to the many authors, publishers, and agents whose permission to quote from their books appears on the copyright pages; and especially, once again, to Gilbert Klingel, whose *Inagua* is still the most perceptive account of island life.

RICHARD PERRY

Northumberland

CONTENTS

ILLUSTRATIONS

LIFE AT THE SEA'S FRONTIERS

1: The Birth of Islands

The sea flows and ebbs, rippling peacefully and beguilingly over the sands, storming violently against the cliffs, building up the land along one stretch of coast, recapturing it along another stretch. In the course of ages she constructs or destroys beaches, deltas, atolls and islands, and in minutes refashions them with her tidal waves or tsunamis. With her tides and breakers she controls all life within her sphere of operations from coral polyp to man, and where the latter engineers vast networks of dykes and barrages to deny her the land, she is capable of overwhelming them; but marvellous are the adaptations with which lowlier forms of life combat the roughest storms, and adjust their lives to the rhythm of the tides.

Consider islands in the sea. They can have been created by separation from a larger land mass, either by erosion or through a rise in the level of the sea, in which case they are continental islands, such as the British Isles or Newfoundland; or they can have been created by volcanic action, either by direct eruption or indirectly through elevation of the sea-bed or by a fall in sea-level, in which case they are oceanic islands such as the Hawaiian Islands or Tristan da Cunha. But the status of some islands is not clear-cut. Madagascar is only 250 miles off south-east Africa, but its restricted and specialised fauna and the depth of the Mozambique Channel, nowhere less than 6,000 feet, both indicate that if it was ever part of continental Africa, it can only have been so many millions of years ago. In subsequent chapters we shall explore the problems of whether the unique Galápagos Islands were created by volcanic eruption and are thus true oceanic islands, or whether they were joined to South America; and whether the Seychelles, which include the only granitic oceanic islands, were not in fact originally a part of a vastly extended India.

We of the television era have been privileged to witness the creation of an oceanic island. It was on the morning of November 14, 1963 that the crew of the *Isleifur*, fishing in 400 feet of water 4 miles off the Westmann Islands, south-west of Iceland, observed a great cloud of steam rising above the sea, and on venturing closer discovered that the sea itself was boiling. By the close of that day the vapour cloud had risen more than 20,000 feet into the atmosphere, and in due course submarine explosions erupted from the surface of the sea in 1,500-feet-high fountains of black ash which gradually established a series of cones above sea-level. These were the foundations of the new island of Surtsey, a little to the east of the position from which a large land mass had erupted 180 years earlier, only to be quickly washed away by the sea. The Westmann Islands had themselves been created by volcanic eruptions perhaps 10,000 years earlier.

Within ten days the Surtsey eruptions had built up a block of lava, some 180 yards wide and more than 600 feet high, resembling an elongated horseshoe, through whose open end rough seas were constantly eroding the new walls of the crater, though instantly vaporized by the hot lava. Arthur Bourne, who visited the island towards the end of November on behalf of *Animals* magazine, has described how massive lava-bombs, forming in the apexes of the ash fountains, were still being hurled 2,000 feet into the air, before crashing down into the sea. For hours at a time the largest of the three craters continued to erupt at intervals of less than a minute, and at night the steep crater slopes of ash glowed red. Yet what impressed Bourne was the eerie quietness of the eruption, for with the exception of sharp cracks and rumbles, emanating from lightning in the vapour cloud, all this enormous explosive energy was being expended in silence.

By December Surtsey was almost 1 mile long, and although the continuing explosions were generally weaker during the course of the winter, there was fresh activity in the spring. By April the volcano had ejected several thousand million cubic feet of ash, and the gap in the rim of the crater had been sealed, preventing the sea from penetrating the active centre. By May 1965, when lava was still flowing intermittently after eighteen months, the island presented a landscape of contorted lava screes alternating with long ash slopes which shelved down to a peaceful lagoon, while beyond its protective bar of cooled lava the cold Atlantic rollers broke impotently on its black beaches, and there were further eruptions between August 1966 and June 1967.

As an illustration of the forces unleashed by a volcanic

eruption, the epic of Krakatau in Indonesia is worth recounting again. Krakatau, now a jungle of trees and shrubs, is situated at the point of intersection of two fissures in the Earth's crust along each of which a chain of volcanoes are ranged. Together with the neighbouring islands of Verlaten and Lang it probably represents the remnants of the crater wall of a colossal volcano which exploded and collapsed several thousand years ago. Subsequent marine eruptions enlarged the Krakatau area of the crater rim to an island, some 6 miles long and 4 wide, itself containing three craters—Rakata (2,623 feet), Danan (1,496 feet) and Perboewatan (399 feet). The latter is known to have erupted in 1680, but thereafter Krakatau slumbered for two centuries until at 10.2 precisely on the morning of August 27, 1883 the greatest explosion in man's recorded history was detonated.

Three years earlier, earthquakes in the region of the Sunda Strait, between Java and Sumatra, had opened fissures in Perboewatan. Sea-water, seeping into the crater, was converted into steam. Pressure built up. And on May 20, 1883 clouds of steam rose high into the sky as the rock plugs of the crater's vents were blown in explosions which were heard 250 miles away. The volcano continued to erupt for the next three months, depositing hot cinders and pumice on passing ships from a mushroom-shaped cloud of steam and smoke, illumined by tongues of fire 3,000 feet high. On August 23

Volcanic Island of Surtsey
—still building up

all was relatively quiet, though there were ominous rumblings from the magma seething below clogged vents; but pressure built up again during the next three days, and at 1 P.M. on August 26 a series of gigantic explosions began, associated with abnormally high and low tides along the coasts of Java and Sumatra. With a still greater explosion at 4.40 A.M. on the 27th, Krakatau began to disintegrate, and thirty minutes later the port of Anjer, more than 30 miles distant in the Sunda Strait was struck by a 33-foot wave. Finally, at 10.2 A.M., with a blast that reverberated 3,000 miles across the Indian Ocean and was heard in Australia, a cubic mile of Krakatau was blown 17 miles high. Pressure waves from the explosion circled the Earth seven times and its dust three times. Half the crater mountain of Rakata collapsed into the chasm, sucking in the sea and then expelling it in a gigantic wave which swept over the port of Merak, 50 miles up the Sunda Strait, in a wall of water 130 feet high.

The remains of Krakatau were peaceful for the next forty-four years, but in December 1927 explosions from a submarine vent 600 feet below the surface presaged an eruption the following year which resulted in the birth of Anak Krakatau—the Child of Krakatau. However, the birth struggles of volcanic islands are prolonged, and Anak was submerged four times beneath the waves, before finally becoming established as an island in its own right in 1930. Two siblings of Surtsey, however, did not survive the stormy wash of northern seas. One, which built up to a length of almost half a mile and a height of more than 300 feet, endured for only five months; the other, slightly smaller, for ten months.

2: Growth of a Coral Reef

In tropical seas volcanic islands are normally fringed with coral, and the extent of volcanic activity during past millennia in the Indian and particularly the Pacific Oceans is beyond conception. There are 2,000 volcanic islands in the Pacific, including Hawaii whose twin 13,000-feet colossi Moana Loa and Moana Kee rise from bases 15,000 feet below the surface of the sea. In the Indian Ocean the 470-miles-long chain of the Maldives alone comprises more than 2,000 coral atolls. Yet three-quarters of the world's atolls are scattered like constellations of stars over the Pacific, and the total area covered by the oceans' coral is perhaps twenty-five times greater than the land mass of the United States.

Although solitary corals are to be found in all the oceans from the Arctic to the Antarctic, colonial reef-building corals can normally only exist in those waters lying between 32°N and 27°S of the equator, where the sea's temperature is not less than 65 degrees F, and actively growing reefs are now found in tropical areas of the mid-Pacific and Indian Oceans, the Caribbean and the Atlantic. Reef corals can also function in regions outside the tropics where these are influenced by warm currents, though they do not develop into large formations. There are, for example, coral reefs in the Gulf Stream off Bermuda, near the mouth of Tokyo Bay at 35°N because of the northerly projection of the warm Kuroshio Current, and below 30°S off Durban because of the southerly penetration of the Mozambique Current. On the other hand there are no reefs on the west coasts of South America and South Africa because these are flanked by the cool Humboldt and Benguela Currents, nor in tropical waters carrying the sediment of large rivers in flood, for corals can only live in pure salt water free of slime, sand or mud which is likely to suffocate them. In the same way, downpours of tropical rain may prove lethal to them at low water, and 100 mile per hour cyclones whip up waves that tear the living coral apart; moreover a muddy bottom does not provide a satisfactory foundation for corals to build on.

The reef-building coral, or madrepore, is a thimble-shaped polyp, resembling the anemone of rock pools but ranging in size only from a pinhead to a pea. One end of its soft body-sac is closed, and the mouth at the other end is fringed with stinging tentacles or nematocysts with which it captures its minute plankton food. Each polyp's jelly-like body is enclosed in a skeleton or theka of almost pure calcium carbonate, which it absorbs from the water and is continually secreting, so that it rests in its own shallow pit or supporting cup of this natural limestone, fashioned to fit its complex folds and hollows. When it dies only the stony skeleton remains. Colonial reef building is made possible by the polyps' ability to grow upon the skeletons of their predecessors, which accumulate in immense compacted masses. As earlier polyps die, succeeding generations, building upon the limestone foundations of the dead, divide and redivide many millions of times and add their strata of limestone, as the colony expands outwards and upwards towards its ceiling near mean tide-level. Thus every colony's millions upon millions of polyps, living or dead, co-operate in creating a communal porous honeycomb structure of limestone. "An entire cosmos, at the level of the infinitely small [as Jacques-Yves Cousteau has conceived it]; millions and billions of tiny beings . . . imprisoned within their own skeletons, who, with their explosive cells and their hooks and their poisons, trap, kill and eat . . . crustaceans, larvae, plankton, and even small fish."

Coral reef with coral polyp

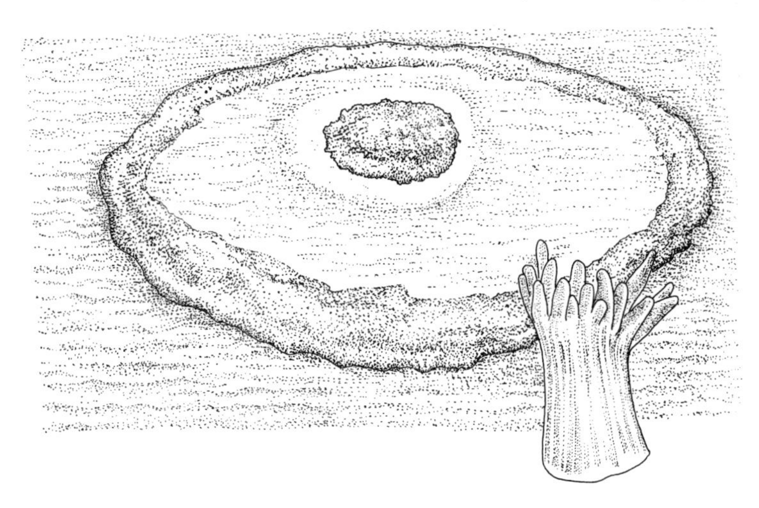

Corals are, however, related both to anemones and to jellyfish, and just as a common British jellyfish passes the winter in a polyp-like form affixed to a rock, before splitting into a series of discs which swim away as individual jellyfish, so the polyps of some corals reverse this procedure and have a free-swimming medusoid form in the beginning. The larvae, from 1 to 2 millimetres long, are ejected from the calices of the polyps, which are mainly hermaphrodite, and swim around for upwards of a month, before settling permanently on some suitable rock to develop into minute polyps, whose stems form star-shaped chambered cups into which they can retreat during the hours of daylight.

There are some 2,500 species of coral, and each fashions a characteristic skeleton, whether it be the delicate blue filigree of the 5-foot spreads of table-coral, or the fragile branching antlers with blue points of staghorn coral, or the rounded boulders and huge clumps, 10 feet high, of the brain-coral, whose convoluted surface is disturbingly suggestive of the human brain. Brain-coral usually grows in association with fantastic jungles of blue, mauve or green palmate, fan or elkhorn coral, *Acropora palmata*—the predominant outer reef coral—often growing to seaward of the main reef. Its interwoven antlered branches form great parasol-like structures of massive girders and cross-beams lying in parallel tiers one above the other, and so large that a diver can swim around under them as if beneath the canopy of a tree. In shallow waters this palmate coral grows to a height of 6 to 8 feet, but in such localities as the Virgin Islands may reach heights of 50 to 60 feet.

Even a single species of coral may exhibit as many as seven different shapes or morphoses—branched perhaps, or elongated, or massed in a ball, according to its position on the reef, the depth at which it is growing, and the nature of the currents. Different species favour different parts of the reef. Some grow in sunny shallows in the lee of the reef, some where the sea surges over the top of the reef, others on offshore reefs in deeper water. In regions such as the Red Sea, where coral grows especially profusely, and undersea cliffs are covered with their thickets, different species are engaged in a constant struggle for space, incessantly losing and recovering territory; but there must, as Cousteau has suggested, be a natural term to this incessant multiplication of collective coral, a saturation point beyond which a colony can multiply no further.

The growth of every colony is controlled mainly by the degree of light penetrating the water and by the strength of the currents, though there is no unanimity of agreement about the various factors involved. The polyps, being sedentary, must necessarily live in flowing water in order that they may be continuously supplied with food organisms; but although corals appear to flourish in the zone of agitated water not far from the surface, the more fragile types tend to favour the quieter deeper waters. Some marine biologists indeed maintain that corals do not thrive in strong currents, that wave action stunts their growth, and that they never grow upwards towards the pounding surf, which continually breaks them down into silt and sand. A contrary view, however, was expressed by the American naturalist, Gilbert Klingel, in describing the Bahamas island of Great Inagua:

> Inagua was growing into the wind. The reef was most luxuriant where the surf was most violent; ever outwards the tentacled polyps were reaching, seeking the swirling water with its billions of micro-organisms, striving to escape the bogging sand of the lagoon behind . . . the trees of coral thrust their

> fingers towards the direction of the greatest current, toward the path of the incoming waves, which is the direction of the rushing trades . . . anchored to their strong homes they cannot pursue their prey; it must be brought to them; that is why coral reefs are always most abundant . . . on the windward side of an island.

The truth is that, although a coral reef is basically composed of the polyps' massive deposition of limestone over thousands or millions of years, a great deal of material is also contributed by the shells of mollusks, the spicules of sponges and sea-fans, and especially by the limy skeletons of nullipores, a family of red algae distributed through the world's oceans. Some large reefs may indeed comprise more algae than coral; moreover, since the former can grow at depths of several hundred feet, they lay foundations on which the true corals can build. Some nullipores, such as *Halmeda*, pack crevices in the coral with their deposits of lime, and also constitute most of the sand on the lee side of a reef. Encrusting algae, like the mauve *Lithothamnion*, growing in breaking seas where no coral can survive, cement the matrix of the reef against the pounding waves; while another, *Porolithon*, may indeed be the deciding factor in making it possible for a coral reef to withstand surf. This alga, with its solid, pink or red limy skeleton, grows over dead coral (though it may also smother the living coral) in areas where the surf is most persistently violent, encrusting the reef with a coating of concrete-like limestone, cementing not only the solid coral but also fragments of it. Near the surface and on windward and seaward sides of reefs only massive forms of coral can withstand the destructive impact of the constantly pounding waves, and in this zone the brunt of the wave shock is borne by these calcareous algae.

What is certain is that the living reef of coral grows upwards and outwards towards light and warmth, and never downwards into the dark cold waters of the deeps; and that most living reefs grow in 90 feet or less of water, and rarely in depths exceeding 150 feet, below which sunlight gradually filters out, though colonies of insignificant dimensions have in fact been found below 500 feet. The upper waters are, of course, where the bulk of their plankton food is concentrated. During the day most of the polyps contract and become inactive; but they expand their fishing tentacles during the hours of darkness, when plankton and swimming worms colliding with the nematocysts are devoured in a few seconds —"tentacles frantically reaching out for food, mouths by the million ingesting minuscule prey, the community eating and digesting as one," in Cousteau's words again. Indeed, some of the so-called "soft" corals, in which the limy skeleton is replaced by stiff spicules, actually increase in size at night from 2½ inches to as much as 16 inches, swelling up into plump pink transparent trees with clearly visible mouths.

It is a curious fact that the polyps of many kinds of corals have green algae living within their tissues. It is known that if plankton food is scarce the polyps actually feed on these single-celled plants, and that some species are indeed no longer plankton feeders, but are wholly dependent for nourishment on these algae. However, it is also a fact that corals, kept in the dark in aquaria and supplied with food, survive although the algae die: whereas those kept in bright light, which the algae require, starve if not fed. It is probable, therefore, that the algae also perform other essential functions, such as the absorption of the polyps' ammoniac and nitrogenous waste products, while Hans Hass has produced strong evidence that the additional supplies of oxygen released by the algae may be a major factor in enabling the polyps to build coral reefs. Whatever may be the precise functions that the algae perform, those corals housing them must necessarily live in the sea's upper 150 feet, where sufficient sunlight penetrates for these green plants to achieve photosynthesis, and where the turbidity from suspended sediments brought down by large rivers does not obscure the light, though it is true that many corals display a remarkable ability to remove silt from their surfaces.

We know very little about the growth rates of coral reefs, though these must vary according to the species and, no doubt, the particular environment. Cousteau noted that a cable, abandoned at his second underwater living station, Conshelf 2, was covered four years later by a growth of palmate coral 8 inches in diameter; and Hass states that while this branched coral can form colonies almost 5 feet in diameter in fifteen years, compact species such as *Meandrina* —which somewhat resembles brain-coral—and the rounded blocks of the finger-coral, *Porites*, achieve maximum growths of only 8 or 10 inches in the same period. The average growth of a single colony is possibly 1 inch a year, but that of the reef as a whole much less, because of such limiting factors as encroachment by nullipores, destruction by predators and storm damage, and amounts perhaps to no more than 18 inches or 3 feet in a century.

The living colony of polyps is attacked by many enemies. It is constantly being eroded by hosts of polychaete boring-worms, especially the palolo which live in burrows in dead coral, and these are probably the prime agents in breaking down reefs. Molluskan snails, cutting into and dissolving the limestone matrix, eventually excavate great caverns and broad overhanging ledges, which provide shelter for the countless inhabitants of coral reefs. Date-mussels bore into the limestone by dissolving it with an acid chemical they excrete, and some *Tridacna* clams (though not the giant clam) burrow into the coral when young and thereafter live within it, enlarging their burrows as they grow. The constant daylong browsing of parrot-fish on the polyps plays an important part in the breakdown of the reefs, for they scrape off not only the polyps but also considerable quantities of limestone which they defecate as clouds of white sand, after from two to eighteen hours' digestion, depending upon the relative size of the individual. Parrot-fish are extremely numerous, and in the Bermudas, where there are reckoned to be a hundred to every acre of some reefs, it has been estimated that they may lay down deposits of this pulverised limestone at a rate of between 1 and 5 tons per acre per year.

All these predators are natural hazards which the coral colonies are adapted to withstand, and from whose ravages they can regenerate; but when man destroys the conches that prey on such starfish as the crown-of-thorns which eat coral, and when he pollutes vast areas of the sea with his oil, sewage and ship waste, then the continued existence of coral is at risk. As recently as 1971 Cousteau reported that the coral reefs at either end of the Red Sea were progressively dying, and that living reefs were only to be found in a very limited area around and south of Port Sudan. In addition, the reefs in the Mozambique Channel were already dead; the large Tulcar reefs south of Madagascar, and those off the Chagos Islands and the Seychelles were decaying; while in the Maldives he noted that thriving communities of marine life were comparatively rare, although coral reefs are normally, as Cousteau describes, the most densely populated region in the ocean, exquisitively coloured by rainbows of tropical fish:

> Triggerfish with vivid blue lines and streaks, their Fernandel eyes eternally curious, staring at butterfly fish and imperial angelfish striped as though by the brush of some colour-mad genius . . . surgeonfish handsomely attired in professional blue, with golden spots on either side of their tails, accompanied by their scalpels—the Moorish idles, whose pennant-like dorsal fins resemble nothing so much as the instrument of a primaeval surgeon . . . swimming among tight banks of yellow snappers and soldier fish.

Many of the coral fish use the ledges, excavated by the predatory borers, as resting places during the day; but at night—when some, such as snappers and grunts, change colour from their brilliant diurnal hues to blotchy gray and white—they disperse over the broad expanses of marine grasses, coral rubble and sand. These local dispersals explain how a coral reef can provide food for such an apparently intolerably high number of fish. By day the reefs belong to the fish, at night to the invertebrates. The American biologist, Robert Schroeder, observed that during the hours of darkness the immense diurnal shoals of fish were not to be found, but only solitary gigantic groupers and snappers peering from grottos in the reef. In the fishes' place black sea-urchins grazed on the algae, the feathery arms of plankton-fishing brittle stars protruded from crevices and openings in encrusting sponges, and thousands of bright little eyes pinpointed the darkness. The majority of the latter belonged to the red coral-shrimps, flaming orange scarlet. Seldom seen by day, they were believed to be rare; but the diver knows that at night they occupy every square foot of ledge on the submerged reefs.

3: The Making of an Atoll

There are three main forms of coral reef—the fringing reef, the barrier reef, and the atoll. The first is a ribbon of coral growing in shallow waters along a coastline or around a tropical island. Its formation can be explained by assuming that, at some time in the distant past, polyps settled in these coastal waters and began to build upwards and outwards in their customary manner, until their colonies reached the low-tide level, where the uppermost polyps could not survive exposure to direct sunlight or, without the cover of nulli-pores, the pounding of the surf. The maximum extent of the colony would have been reached near the surface, and have decreased gradually in size downwards to the 150 feet limit. As the reef expanded outwards a shallow trough would have come into being between it and the coast, and this would be the receptacle for detritus from both coast and reef, and also for the silt carried down by rivers. In due course this debris would smother and kill the polyps on the landward face of the reef, and the shallow channel between coast and reef would become permanent.

Fringing reefs and barrier reefs can lie adjacent. In the Red Sea the 300-miles-long and 30-miles-wide Farasan Archipelago comprises a fringing reef separated by a narrow channel from a barrier reef. The latter differs from the fringing reef in that it rises steeply from the sea-bed in deeper water, and is therefore separated from the mainland by a broad channel or by fringing reefs. The coral complex of Farasan is second only in area to the Great Barrier Reef, which stretches for 1,260 miles from Cape York almost to Brisbane as a colossal natural limestone breakwater of polyp skeletons, lying from less than 10 to more than 200 miles off the Queensland coast. But for only half this distance, between Cape York and Cairns, does it form an unbroken rampart, about 180 feet thick, against the great white Pacific breakers which here and there have piled up high white mounds of sun-bleached shells—cowries, pearly nautilus, spider-shells, cat's eyes and pencil shells. Elsewhere in the Coral Sea between the coast and the Outer Barrier, which rises precipitately from depths as great as 4,600 feet, it comprises a jumbled and irregular series of reefs in long, more or less parallel serrations, some lying beneath the surface, others above water and covered with sand. Many are submerged at high water but emerge for one and a half hours at low tide from the long blue swells of the Coral Sea to reveal (as the Australian naturalist, Graham Pizzey, has described) endless forests of coral inset with turquoise pools and gorges where shafts of sunlight spear through motes of drifting fish to the dazzling bottom perhaps 20 feet below. In the gorges leather-flippered green turtles flap through the clouds of fish, and cast their shadows over pale blue stingrays lying in the sand and painted crayfish, or spiny lobsters, crouching under coral shelves. By far the most numerous fish are the small oval and laterally compressed *Chaetodontids*, which display an extraordinary range of colours and patterns among their 150-odd species. None are more fantastic than the angels and butterflies with their distinctively pointed, multi-toothed mouths adapted to picking invertebrates from cracks in the coral and rock.

In the pools are pastel-coloured gardens of coral and sponge inhabited by the same clouds of drifting fish, by prawns with tiger stripes, by baler shells and cone shells equipped with hypodermic needles loaded with a possibly lethal nerve poison, and by giant clams measuring from 4 to 6 feet and weighing as much as 400 pounds. When the latter open their shells, in order to pump plankton-bearing water through their siphons, they extrude fleshy mantles, coloured white or green, brown or vivid red, and from clear blue to purple or black, by colonies of algae which provide the clams with vegetable food. The giant clam is one of the very many marine creatures about whose habits the evidence is contradictory. The popular belief that a diver could be drowned by being trapped by the foot in a clam's gape seemed to have been confuted when it appeared that the mollusc required from 30 to 60 seconds to withdraw its mantle and close up; but an American team of scientists, while investigating the general ecology of Ifaluk, an atoll in the western Caroline Islands, reported that in at any rate some species the shell shuts with remarkable speed, because the closing mechanism is activated by a form of primitive electric-eye. Hundreds of bright eye-spots on the extruded mantle are in fact transparent cells that focus light on to tissues below the skin; and when the light is interrupted by the shadow of an object, the shell instantly snaps shut.

However, there is, I believe, no indisputable record of a diver being trapped by a clam.

The third major form of coral reef is the atoll. There are still various hypotheses about the origins of atolls, and it is unlikely that all have been formed in the same way, in view of the fact that coral reefs are engineered by so many different kinds of polyps in a variety of environments. But, first, what are the outward and visible signs of an atoll? Francis D. Ommanney, whose fishery researches have taken him to a delectable variety of seas from the tropics to the Antarctic, has given us a general picture of an atoll:

> When you have seen one, you have seen them all. The sea breaks in a white line around each one, leaving between the reef and the shore a shallow lagoon of opalescent green. The shore is glaring white in the hot sun and lined along its landward edge by *Scaevola* bushes. Then there are a few ranks of feathery casuarina trees, and for the rest, rank upon rank of coconut palms in beautiful and negligent orderliness . . . leaning gracefully this way and that, their elegant lichened trunks of elephant grey expressing the langour of the hot climate, their stiff dark plumes, bursts of metallic green, rattling and whistling as they bend to the trade winds.

An atoll, then, is a living and growing chaplet or bracelet of coral reefs, often hundreds of miles from true land and rising abruptly from mid-ocean depths of several thousand fathoms. (Rocas Reef, 135 miles off Brazil, is the only one in

Under water among the corals

Gahan

the Atlantic.) The reefs may be almost submerged or rise partly above water to form a low island on their windward side, crowned with coconut palms or covered with scrub which, in some wet regions, may be luxuriant. On the inner rim of the reef, from 300 to 500 feet broad, sands slope down to an, in theory, but by no means always in fact, approximately circular lagoon of almost enclosed quiet ocean waters. The lagoon may be a mile or less in diameter or perhaps several miles, the largest probably being Kwajalein, in the Marshall Islands, which is 15 nautical miles wide and 63 miles long—a sea within a sea.

Volcanic isles are, as we have seen, usually fringed with rich growths of coral thriving on plankton brought to them by currents, and the great majority of atolls are certainly based on volcanoes which have sunk beneath the sea because of weathering or erosion, because of a rise in sea-level due to the melting of glaciers, or because subsidence in the sea-bed has led to a volcano becoming either a sea-mount with a peak reaching into the upper sunlit waters in which corals grow, or into an island protruding above the surface. If a volcanic island begins to subside, the colonies of coral polyps and other reef-building organisms continue to grow upward on the surf-washed outer edge of their ever-widening reef. Their slow growth keeps pace with the very gradual subsidence, or for that matter with a rise in the level of the sea, and what was in the beginning a fringing reef around the island is transformed, as more and more of the island sinks beneath the surface, into a barrier reef with a zone of open water between it and the island. Ultimately the reef almost completely encircles the crater of what was once a high central volcano and, when this is submerged, the reef remains as a submarine coronet of coral around the drowned apex.

In the final stages of an atoll's formation the sea wears away the reef-ring and floods in through the gaps it has eroded, breaking the reef up into smaller segments and forming a chain of islets around a shallow lagoon, in which the original volcanic mass is deeply buried.

Since it is the coral on the outside of the reef that grows fastest in the surf-bearing plankton, until killed by the sunlight when rising above the surface, the reef does not grow over the submerged crater. Hence the lagoon which, since it is not scoured by waves and currents, becomes partly filled with sediments and with the polyps and skeletons of certain lagoon-growing corals, and remains comparatively shallow. Exceptionally, indeed, a lagoon may be completely filled in, as has been the case on Europa Island in the Mozambique Channel. On the other hand, a lagoon can itself be instrumental in creating new atolls. There are, as we have seen, more than 2,000 of these in the Maldives. They rise from a submarine plateau lying at a depth of about 130 feet, and are in fact composed of ten enormous reef-circles, some as much as 40 miles in diameter, which contain a multitude of very small atolls or faros. Hans Hass, who was the first man to skin-dive among the Maldives, observed that these large reef-circles and tiny faros looked as if they had been constructed by someone sitting on the sea-bed and blowing large and small smoke-rings up to the surface. He discovered that the interiors of the reefs he examined were not cemented together and solid, but porous and unstable. He therefore conjectured—if I interpret his somewhat involved account correctly—that the curious formation of these Maldive atolls was due to the fact that their unstable reefs had sagged on their inner sides where there was least coral growth, thereby increasing the depth of the lagoon. At the same time the sea, breaking channels through the reef-ring in many places, has poured through these into the lagoon, and not only assisted in deepening the latter still further by breaking down the dead coral, but has created suitable conditions within the lagoon for the growth of new coral reefs, and the subsequent building up of new small atolls.

The construction of an atoll is measured in millennia. More than 4,000 feet of coral limestone—the product of perhaps 50 million years' engineering—caps the basalt foundations of Eniwetok Atoll's sunken volcano with what is apparently a hollow cone of hard limestone filled with loose sediments. Bikini Atoll rests on no fewer than eight volcanic peaks, and it is estimated that an average-sized atoll contains 15,000 times more material, in the form of coral, than Egypt's largest pyramid. If, in a few hundred thousand years, there is still pure unpolluted water in the oreans, Bora Bora will be an atoll. Bora Bora, one of the Society Islands, is reputed to be the most beautiful island in the world, with its reddish-black volcanic rock and green slopes, its palm-capped reef of sand fringing the greenish-blue lagoon, and its surrounding deep-blue sea. But it is a very ancient, sinking island. All that remains of the rim of its immense crater is a long low rectangular ridge, the existing island's backbone, and an adjacent conical one.

4: Life Comes to an Atoll

The foundations of an atoll are the coral reefs surrounding the central lagoon. On these foundations must be built a crown rising a few feet above the surface of the sea. The reef grows outward towards the life-giving surf. As it broadens outwards so its mid portion is no longer within the zone of growth—as is the case with the landward side of a fringing reef. The dead limestone is eroded and breaks down into sand, and the larger the reef-circle the greater the area of the interior sand laid down in this way. Winds, and the tides that stream to and fro through the channels in the reefs, heap up the sand into shifting banks. All now depends on at least a part of such a sandbank remaining stable sufficiently long for plants to become established, spread over the whole bank, and bind the soil into a structure that can withstand the pounding of the waves. In some instances, as the soil is stabilized and the vegetation increases, so the extent of the central lagoon decreases, and what began as the formation of a reef by the subsidence of a volcanic island becomes an island by the upward growth of an atoll.

An atoll can be crowned in another way. At Addu Atoll in the Maldives, Hans Hass discovered that large reefs had broad platforms, 6 to 12 feet below the surface, which were largely overgrown with table-shaped palmate coral. When this coral died, the platforms had been turned over by the waves and had subsequently become so thickly encrusted by calcareous algae that their delicate ramified branches had been transformed into solid slabs of stone. Storms and tidal currents had piled up loose rubble above the mean tide-level, and surplus fragments had been swept over the reef on to its leeward side to form a pile of debris on which new coral colonies could not grow, but which would provide the foundations for an islet. Some of the Great Barrier Reef's sand-cays appear to have been built up on submarine masses of such reef debris 400 feet or more thick.

The creation and retention of a crowned atoll must owe much to a combination of favourable circumstances, for the highest elevation of islets based on atoll reefs does not normally exceed 15 or 20 feet, and many islets lie only a foot or two above high-tide level. None of the Maldives' now palm-clad atolls, for example, rise more than 6 feet above the surface of the sea. The most remarkable fact about the colonization of atolls is surely not that a flora and fauna may eventually become established on them, but that they are not repeatedly devastated and swept away by stormy seas, or inundated by tsunamis against which virtually all are defenceless. However, apparently few atolls are struck more than once in a quarter of a century, at worst, by tsunamis.

Popularly known as tidal waves, though erroneously since

they are not the product of tidal forces, tsunamis are engineered by earthquakes in the sea-bed and may travel as much as 5,000 miles from their place of origin in waves extending from 100 to 600 miles from crest to crest—in contrast to the 320 feet between the crest of Atlantic waves or the 1,000 feet between those of Pacific rollers—and at a velocity of 500 miles per hour in the deep waters of the Pacific in comparison with the 60 miles per hour of normal storm waves. For many hours the tsunamis travel in trains of waves, fifteen minutes or more than an hour apart because of their great length, and so shallow in comparison with their length that they are hardly detectable in the open ocean, being no higher than perhaps 2 feet from trough to crest. Only when nearing coastal shoals or flat shores do they build up to heights of from 20 to 60 feet, and to mountainous proportions when pouring into V-shaped inlets or harbours. Again, the initial rather sharp swell is little more threatening than an ordinary wave and affords no indication of what is to follow, until a tremendous suck of water away from the shore, leaving reefs high and dry and multitudes of stranded fish, heralds the arrival of the first great trough, and then a succession of colossal waves, of which the third and eighth are usually the largest. Great numbers of deep-water fish may be brought to the surface; small fish, such as sardines, may be distended with the quantities of disturbed diatoms on which they have gorged; and at night the sea may glow brilliantly with the luminescence of the dinoflagellate *Noctiluca*, stimulated by the turbulence.

The lime basis of broken-down coral only requires the addition of a little windblown sand or sea-borne mud to form a fertile soil in which seeds can germinate and strike root. New volcanic islands are so thoroughly burned and so thickly covered with ash and pumice that it appears inconceivable that they can ever support life. Yet the very substance of a submarine volcanic eruption contains the essential ingredients necessary to nourish life. Only a few days after Surtsey had risen above the waves, ash was removed to the Field Studies Centre in Surrey and potted. Within two months, common moss had germinated in the pots, despite the fact that these had been covered by 3 inches of snow for one week, and after the moss, various grasses and, within eight months, common groundsel. Moreover, even while Surtsey's volcano continued to erupt and intermittently cover the island with hot ash and lava, organic material, washed up by the sea and absorbed into grains of sand along the tideline, provided a nutrient layer for bacteria, including one similar to a type previously recorded only from Loch Ewe on the north-west coast of Scotland 650 miles distant.

Sand-cays, islands and archipelagoes lying a score or two of miles offshore can obviously be colonized without difficulty by plants and animals from the adjacent mainland, while those islands that have been sundered from the mainland by erosion or by a rise in the level of the sea will already contain samples of the original continental flora and fauna. In those 80,000 square miles of the Coral Sea's inner waters enclosed within the Great Barrier Reef are many thousands of shoals, reefs, atolls, coral cays and islands, 600 of which are large enough to have been named, such as the 140-acre coral cay of Heron Island and 30-miles-long Hinchinbrook with mountain peaks rising to a height of 3,000 feet. There are also high islands (the summits of drowned mountains), of which the larger ones' flora is similar to that of the mainland and includes such trees as hoop-pine, magnolia, flame tree, wild plum and eucalyptus. In addition a chain of untypical islands have been separated from the Queensland mainland by subsidence or by a rise in sea-level. Dunk Island, for example, which is composed of volcanic rock, whose red soil supports grasslands and a dense rain-forest rising to its 900-foot ridge, contains much wildlife not found on the coral cays.

But how are atolls and islands in mid-ocean, hundreds or thousands of miles from the nearest land, to be colonized? In their isolation they can be likened to a snow-capped mountain, such as Kilimanjaro, towering from a tropical forest, or to an oasis in the desert, though the oceanic island is the most isolated because the sea forms the greatest barrier to the migration of terrestrial fauna. Its plant and animal communities are therefore derived from the few species able to traverse extensive areas of landless ocean. Seeds and fruits for these new sea-lands must either float to them or be carried by some agency, and in either case be in a suitable condition to germinate on arrival. There are four possible carriers—birds, winds, sea currents, and man. Birds and winds must be the main agents because few seeds and fruits could survive long passages by sea except in men's canoes or ships. Birds, and sea-birds in particular, are ever ranging over the oceans, whether on migration or fishing or, like the albatrosses, just wandering with the westerlies of the "roaring forties." Guillemots, puffins, gannets, fulmars, kittiwakes, and larger gulls habitually rest on Rockall, almost 200 miles west of St. Kilda and the most isolated small rock, or the smallest isolated rock, in all the world's oceans. A gannet may alight on its 70-foot summit with a beakful of oarweed, and a score or two of guillemots may crouch on a ledge with their backs to the sea as if incubating eggs, but it is highly improbable that any nestling sea-bird was ever hatched on this Atlantic stack over which white water spouts to a height of 170 feet.

Two herring gulls touched down on Surtsey as early as the

second week of its emergence. The following spring a number of redwing thrushes and a single snow bunting rested on the island while no doubt en route to their nesting grounds in Iceland. Dunlin, oystercatchers and kittiwakes were the next visitors to be recorded, and by 1965 twenty-three species had been identified. Surtsey ceased to be active early in June 1967, and today large numbers of birds alight on Surtsey during their spring and autumn migrations, while kittiwakes form very large summer roosts and as many as 10,000 larger gulls, especially immature greater black-backs, gather at the lagoon and on a beach on the north coast. Their droppings in the lagoon nourish algae and rare flagellates.

Birds play an important role not only in introducing vegetation to an atoll but also in building up its soil and in influencing the ecology of the surrounding seas. Though they prey on the fish and other marine fauna in the waters around their breeding stations, their dung provides food for the lower life and enriches the water for plankton. What they take from the sea they add to the land in organic material, altering an atoll's soil and vegetation. A flight of sea-birds alights to rest and roost on the coral debris of an atoll, observes Graham Pizzey. Their excrement, rich in phosphates, provides additional fertilizers to assist the germination of grass seeds and small herbs. The plant-stabilized coral pile then traps more debris, more flocks of sea-birds alight, and finally colonial species such as sooty terns establish breeding colonies and manure the surface of the atoll, generation after generation, with guano. Its crown is overlaid by a nutritious grey soil composed of guano and broken-down coral grit, enriched by a compost of dead birds, food remains, flotsam and rotting bird-introduced vegetation. The terns have begun the initial conversion of a pile of dead coral into what may prove to be a forested marine oasis of wild life.

The prolonged accumulation of guano even on quite small coral cays such as the islands of the Lacepedes off north-west Australia is hardly credible. Caspian and roseate terns, pelicans, brown gannets and lesser frigate-birds (pirating the gannets) nest on these cays, and in a single year in the late nineteenth century more than 37,000 tons of guano were mined there. When the mine closed down, the open-cast hollows were recolonized by the gannets.

These immense colonies of breeding sea-birds favour the drier barren atolls, undeterred by the absence of fresh water since they obtain sufficient liquid from the fish they eat, and can expel any surplus of salt through the medium of special nasal glands able to extract a highly concentrated solution of salt from the blood. Possibly their preference for this type of atoll is associated with the general absence of predators, crabs excepted, though on the Lacepedes the nesting birds are preyed on by black rats, which probably landed initially from pearl-hunters' luggers stranded on the mudflats and now inhabit burrows among the spinifex scrub, while skinks (lizards) are extremely numerous on some atolls and take a heavy toll of both eggs and newly hatched nestlings. On oceanic islands relatively unaffected by man, by his introduced rats and cats, and his commercial exploitation of eggs, plumes and guano, sea-bird colonies can extend to the limits of the space available and exploit this to the maximum with extraordinary ingenuity. On Gough Island, for example, lying 230 miles to the south of Tristan da Cunha, the millions of sea-birds have divided up this volcanic island, which is only some 8 miles long by 4 wide, into specific breeding zones. Rockhopper penguins, Antarctic terns, great skuas and the world's only breeding colony of 4 or 5 million great shearwaters occupy the coastal zone; the sooty shearwaters, grey-backed storm petrels and yellow-nosed albatrosses the wooded and tussac zone; the Atlantic petrels and pediunkers the grassland; and the wandering albatrosses the boggy plateaux. Moreover, the various species phase their seasons of arrival at and departure from the island so that one species takes over a zone in which another has concluded its breeding cycle, with the result that the island is tenanted by the various breeding populations all the year round.

Sooty terns, probably the most numerous and therefore the most successful of tropical sea-birds, have evolved a system whereby their breeding seasons in different localities coincide with the maximum supplies of fish for their chicks. One of the largest colonies of these terns is to be found on Desnoeufs, one of the Amirantes some 200 miles from Mahé, the main island of the Seychelles, though commercial egg-taking has reduced its strength from an estimated 5 million in 1931 to less than 1 million today. Also nesting on Desnoeufs are some thousands of noddy terns, and since the island's 86 acres rises only 10 feet above the sea at its highest point and is only ½ mile wide, every open space is colonized, while wedge-tailed shearwaters occupy burrows in the friable soil beneath the colonies of terns. Indeed, the sooty terns fortuitously appear to enlarge the potential nesting area. There are only a few stunted palms on Desnoeufs, together with some coarse grass and a low-growing scrub bush in the form of a ground creeper, in which many of the nesting terns become entangled. The noddies lay their eggs on rocky outcrops and the sooty terns in places where the vegetation is dead or absent. K. B. Newman, who visited Desnoeufs some ten years ago, observed that the multitudes of terns filling the air over the island to a height of 300 feet, continually sprinkled the vegetation with a light "rain" of salt water from their nostrils; and he noted, further, that once they had established their nesting colonies, the vegetation died off very quickly, with the result that later in the season their colonies were larger.

But none of these hundreds of thousands of breeding birds attempt to nest until the onset of the south-east monsoon about the middle of May. This is associated with an upwelling of nutrient salts from deep waters off the island, which results in increased supplies of plankton for the young fish, spawned at this appropriate season when they provide food for the nestling terns. On Desnoeufs, and throughout most of their range, seasonal variations in fish supplies make it possible for the sooty terns to breed annually in the normal way; but on Ascension Island in the south Atlantic they are obliged to nest at intervals of 9½ months at different seasons in successive years, while on Christmas Island in the central Pacific there are two separate six-month nesting periods each year, with those breeding successfully during the first period not returning until the following year, but those that are unsuccessful returning again in the second period. It is not surprising that sooty terns are numerous.

Although there may have been a tendency to exaggerate the numbers of plants, shrubs and trees introduced by birds to virgin atolls and islands, because their ability to do so depends in a proportion of cases on the time that elapses during an oceanic flight before a bird excretes the seeds or fruits it ate at its last port of call, there is no doubting their potentialities as carriers. Darwin made a number of experiments to appraise these, and in a more recent one Gilbert Klingel, having rinsed enough mud to cover an area the size of his little finger-nail from the feet and legs of sixteen spotted sandpipers, which had migrated to Inagua, separated from this soil eleven seeds visible to the naked eye; while from the corpse of a little green heron whose legs were caked with mud he removed seventy-eight seeds with downy strands of whitish silk similar to the fluff of cat-tails.

> When one considers [he commented in his enthralling island book *Inagua*] the vast number of sandpipers and other birds which each year sweep down through the Bahamas on the winter migration from North America, and consider the tremendous number of seeds, spores, microscopic algaes and one-celled animals that must be carried on their feet, tucked in crevices between their claws, under scales or clinging to their bills, the wonder is not that there are so many animals, and plants, on Inagua, but that there are so few.

Krakatau was only a dozen miles from Sibesia, the nearest land not devastated by its eruption, and within 9 months a few blades of grass were growing on its charred remains, within 3 years at least 15 flowering plants and 11 ferns, and after 50 years 271 plants. Its colonization appears to have been more successful than that of its child, Anak, for though 20 species of shore plants were growing on the latter within 2 years of its ultimate establishment as an island, all were destroyed the following year, and this rhythm of alternate colonization and destruction has continued up to the present time. Similarly, the first seeds and plants—sea-rocket, lyme-grass and angelica—that drifted ashore on Surtsey in the spring following its emergence failed to strike root, while the first plants to do so subsequently, those of the sea-rocket, were either buried under volcanic ash or were destroyed by storm waves. Estimates as to how Krakatau's plants reached it vary wildly, but if the most recent estimate is the most correct, then perhaps rather more than a quarter arrived with sea currents, a quarter were carried by birds, and about 40 per cent were wind-borne. To oceanic islands such as Gough, more than 1,800 miles from Africa and more than 2,000 from South America, the proportion of wind-borne seeds is likely to be greater. There are in fact only about forty flowering plants on Gough—less than one-tenth the number on barren Tierra del Fuego—and of these those with spores or light seeds predominate, which suggests wind transport.

It is the winds, of course, that carry the majority of insects to islands and, for that matter, influence the geographical distribution of seed-carrying sea-birds and no doubt of some small passerines also, whirled out of their known orbit in hurricanes. For example, the nearest relatives of one of the Inaguan humming-birds inhabit the mountains of Costa Rica and western Panama, and Klingel has suggested the possibility that these humming-birds owe their presence on Inagua to the phenomenon that over Central America, Colombia and Venezuela, rising thermals create partial vacuums as they mount skywards, which the south-easterly trades rush in to fill, sucking up humming-birds and other small passerines.

In one of his innumerable experiments William Beebe, using his ship as a floating research station when 60 miles from the Cocos Islands, was astonished at the numbers of insects and birds (and also plants and shore crabs) that entered the research area during a period of only a few days. In the spring following the eruption of Surtsey three species of insects drifted into it with the wind and, despite the absence of any vegetation, had established themselves as breeding species on the decomposing remains of fish and birds within five years, together with a fourth species which may have been propagating in the tide-rows of rotting seaweed. By this date, 1968, a spider—believed to be a North American species able, when immature, to balloon great distances suspended on silken threads—had also arrived, as had two species of widely migratory moths, the dark sword grass (*Agrotis ypsilon*) and the silver Y (*Plusia gamma*). Neither of the latter are native to Iceland, and were therefore probably migrants from the British Isles or Europe. In like manner at least one spider had optimistically spun its hunting net on Krakatau within nine months of its eruption, and after five years the island's population of spiders, beetles and butterflies was considerable. "Aerial plankton," in the form of insects and spiders, are borne by the winds and upper air currents to all parts of the Pacific, and two species of butterflies with strong powers of flight have spread throughout Micronesia. Nevertheless, many groups of insects have failed to colonize oceanic islands because these lack those plants on which their larvae feed, while others such as dragonflies are prevented from doing so because of the absence of fresh water—frequently scarce on atolls and oceanic islands—essential to their aquatic nymphs. However, some dragonflies have colonized some Micronesian islands; and although some species of migrant Brazilian moths are unable to establish themselves on Tristan da Cunha because of the absence of their food plants, some species of moths must have done so long ago, for today there are wingless moths in the crater of this remote island's 6,760-foot volcano. That they are wingless is possibly a protective adaptation against being blown off this storm-girt island, though it would seem a remarkable feat for a winged moth to arrive on an island and then, in the course of ages, lose its wings by natural selection or some favourable accidental mutation.

We have now to consider colonization by sea. Klingel has described how when the tide was beginning to flow, and tiny ripples were lapping on the beach of an Inaguan sand-cay, whose only vegetation was a single great mangrove tree, he saw rolling in these ripples a long spear-like shaft of reddish wood about a foot in length. This was the radicle of a mangrove seedling which had been washed off some mainland shore and had drifted with the tide and trade wind to Sheep Cay. "Whether it would survive after having arrived was a moot question. But not twenty feet away was another radicle half buried in the sand. From the top of this was emerging a pair of green . . . leaves. Beneath the soil long fibrous roots had anchored themselves, firmly intertwining the wreck of a long buried conch."

That mangroves can survive trans-oceanic crossings is suggested by the fact that those of West Africa are of the same species as those on the east coast of South America, whereas those of East Africa are more commonly allied to those of Asian seaboards; and though one usually associates them with coastal and estuarine swamps they are in fact among the first trees to colonize coral reefs and atolls. There are several species of this remarkable tree, but all possess a unique method of propagation, varying in technique from one to another. Growing in ever-shifting mud or sand, and periodically flooded by tides or pounded by waves, the mangrove is compensated for this insecure environment—as Klingel has expressed it—by retaining the product of its blossoms until these have produced a long stiff radicle or spear-like root suspended from the branch. Two options are now open to the root. It can either continue to grow downwards until it reaches the ground, to which it attaches itself by rootlets and, breaking loose from its parent, begins life as a separate tree; or it can suddenly part from the branch and plunge down like a javelin to bury itself upright in the mud until its roots take hold. Those of the black mangrove proliferate and send up hundreds of little schnorkels or pneumatophores, resembling spikes of asparagus and pierced with minute holes through which oxygen is taken in and carbon dioxide expelled. Thus a single mangrove may in a few years infiltrate many square yards of brackish mudflats. The great red mangrove in the centre of Sheep Cay may have originated from one radicle or from a number, which subsequently coalesced into several hundred separate yet united trunks, supported by a few thousand stilt-like prop roots on a platform raised well above the level of the salt water and covering at least half an acre. "But for this arboreal giant [to quote Klingel again] Sheep Cay would be nothing more than a bar of white, gleaming sand, for ever shifting, drifting at the mercy of the wind and currents. Only under the branches was there any hint of solidarity: here the calcareous sand, held tightly by the binding roots, had settled and metamorphosed into grey carbonate of lime."

The mangroves therefore build up, consolidate and increase the area of islet or atoll, and also offer considerable resistance to hurricanes enabling other trees, such as casuarinas and pandanus and coconut palms, to become established; the delicate foliage of the casuarinas contrasting with the pandanus' stiff clumps of lanceolated leaf-plumes streaming in the trade winds. Although the fruits of the pandanus can survive prolonged immersion in salt water, the small cones of the casuarina must be at least partially dried before they will float and drift on to the sand-cays as ripe and fertilized fruits, and despite claims that coconuts can float at sea for years and still germinate if washed ashore, it seems more probable that in fact they cannot survive for long periods in the sea. Thor Heyerdahl gave particular attention to this detail on various voyages and noted that though coconuts retained their buoyancy for long periods in sea water, this gradually penetrated the region round the eyes with the result that, after two months, mico-organisms affected and ultimately destroyed the nuts' germinating organs. At a guess, a voyage of 300 miles would be about the longest a coconut could achieve and remain fertile, and the majority of coconut palms, which encircle the globe between 15°S and 15°N of the equator must have been distributed throughout their countless islands and atolls by man. Once established, the pandanus palm or screw pine—whose root system somewhat resembles the mangrove's, with a maze of aerial stilt-like roots leaning up against its base like a pile of faggots and propping up the weirdly contorted branches—can grow in shifting sand or mud with little or no access to fresh water; but though coconuts can survive without fresh water, they thrive better on the wetter islands, where their very dense and spongy root system enables them to soak up quantities of water. Even so, they are slow to mature, reaching heights of 70 or 80 feet only after perhaps a century's growth.

Although many of the Pacific atolls receive 100 inches or more of rain a year and, in spite of poor soils, may, like Ifaluk, be covered with a lush cap of greenery from beach to beach, drier atolls may have strictly seasonal rainfall and be subject to months or years of severe drought. Indeed, near-desert conditions may persist on the driest atolls, with the result that the flora on some, such as the northern Marshall Islands, may be restricted to as few as nine species of plants. But on any atoll fresh surface-water is scarce or non-existent, because any rain that falls is almost instantly absorbed into the porous limestone soil, through which it trickles down until reaching the subterranean layer saturated by sea-water. There it gradually collects into a vast pool which, being slightly lighter than the underlying sea-water, actually floats on top of the latter, though pressing it down in the middle and out at the sides in the form of a lens, and mixes with it only very slightly, despite the fact that the levels in an atoll's freshwater wells rise and fall with the tides.

In addition to the mangroves, palms and casuarinas, the heavy, close-grained and iron-hard pemphis thorn thrives on the bare rock, even though this may be submerged at high tide and its plant growth otherwise restricted to algae. The pemphis roots creep down the shore and out to sea, coiling naked reddish-brown tendrils among the boulders on the flat of the reef; and after a typhoon has stripped all the vegetation from an atoll, a pemphis, only 2 or 3 inches in diameter and with half its bark torn off, may be seen sprouting new leaves and budding new branches from its slender stem. But though the latter is normally only a few inches in diameter and rarely higher than 20 feet, the pemphis often forms a narrow but dense strand of smooth gray trunks and sage-green canopies between the mangroves and the coconut and pandanus groves.

Behind this peripheral protective barrier against the storms a forested area can become established, and reclaim the coral cay. Noddy terns alight on an atoll, bringing with them the sticky seeds of pisonias from their roosting and breeding stations on older cays, and compacted thickets of these spring up. Pigeons, which habitually roost in large numbers on coral islets, excrete seeds and fruit stones transported in their crops from other, forested islands. The variety of trees is increased by larger species of pigeons, whose beaks are strong enough to crack seeds which can, otherwise, be broken only by a sharp blow with a stone. Pigeons, silver-eyes and especially spectacled flying-foxes—several dozen of which may establish a permanent roost in a grove of tall erythinas—continue to arrive with fruit seeds from the mainland and make further contributions to the by now heavy stands of trees. The vast mangrove swamps of tropical Queensland shelter the bulk of a huge flying-fox population, whose members roost in the swamps by day, and disperse by night to feed on figs and flowering eucalypts. Eventually there may be fifty species of birds breeding on one small island, which has become so heavily timbered that the sooty terns, who were the original fertilizing agents, are obliged to abandon their breeding station and establish a new one on a virgin atoll of coral and sand.

5: The Ways of Crabs

An atoll's pioneer colonists are sea-birds, insects and crabs. But once even a single mangrove or palm tree has established itself a new environment is in the making, suitable for land birds, lizards and land crabs; indeed, even a solitary bush on a sand-cay may provide sufficient cover for the nucleus of a small terrestrial community. A coconut cast ashore on a coral cay, might bring the first of these new colonists, for its buoyant casings enable it to ride high in the water and thus provide a lizard passenger with a good chance of surviving a short voyage, whereas a lizard awash on a mangrove radicle would soon become chilled and drown. It is conceivable that a mainland bush, containing a family of lizards, could be torn up by the roots and blown out to sea in a storm. Other lizards might be cast adrift on tangled masses of trees floated out to sea in river floods. Large rafts, composed of masses of debris carrying palm trees, 20 or 30 feet high and still erect, are frequently borne westwards from the Fiji Islands; but these rafts do not usually carry lizards and must, in any case, soon be broken up by rough seas. Nevertheless, lizards are great travellers and have colonized much of Polynesia, and one must assume that the majority of lizards on those far-flung islands and atolls can only have reached them in the ocean-going canoes of the maritime Polynesian peoples.

Inagua, only a few miles from a chain of island neighbours, would of course have been a relatively accessible island for any rafting lizards, and they have been there long enough to establish themselves as a peculiar arboreal race varying in colour from pale gray suffused with a tint of lavender to rich chocolate-brown, and capable of rapid colour changes in which they blush yellow or pinkish or greenish-gray. On Sheep Cay the lizards' whole world was centred in the huge mangrove, as indeed was that of all life on the cay; and Klingel observed that the lizards' habitat was in fact restricted to the branches of the tree, whose barrier of leathery leaves not only afforded them shelter against the persistent gale of the trade winds (which may blow without cessation for weeks at a time over Inagua), but also provided that quiet atmosphere essential to the existence of the mosquitoes and other small insects which were the lizards' source of food. Even the land crabs depended on the mud and debris around the tree's innumerable roots to provide them with the microscopic tidbits that they picked up among the dead leaves, and did not venture outside the mangrove's extensive shade into the hot sun.

Sea-birds excepted, the dominant life, both in abundance and ubiquity, on atolls and islets is crustacean—burrowing crabs, fiddler crabs, soldier crabs, hermit crabs and rock crabs, land crabs and robber crabs. Not an inch of the atoll world is untouched by their questing claws, and they must profoundly influence its vegetational complex, for though the hermit crabs are omnivorous and, like the robber and land crabs, useful scavengers, they may seriously denude the vegetation if unable to obtain sufficient animal matter, dead or alive. Hermit crabs may indeed be so numerous on Pacific atolls as to overrun them, and take a serious toll of nestling terns, despite the fact that as they grow larger they must suffer from an acute housing shortage, because of the lack of a sufficient supply of large mollusk shells. Those whose increasing size compels them to move house must either find the comparatively rare cats-eye shell, which measures up to about 3 inches in diameter, or die; yet this predicament does not appear to have a controlling effect on their numbers. However, they tend to be scarce in the central parts of large islets because, not being fully adapted to a terrestrial environment, they must live within range of water in which they can wet their gills. Every evening at sundown they migrate to the shores of the lagoon on Canton Atoll in order to replenish the reserve of water in their shell houses, the interiors of which are gripped so tightly by little hooks at the end of their tails that they appear to be cemented to their shells, although this operation, taking only a few minutes, may involve them in a journey of as much as 200 yards.

So too, where coral reefs are associated with tidal flats of sand or mud, covered perhaps with marine grasses, there are occasions when the whole area seems to get up and move rapidly away, for countless small brown crabs—soldiers or signallers or sand-bubblers—inhabit burrows in the mud and, as the tide ebbs, each pushes its way up through the fine silt blocking its hole. If not alarmed, these tiny *Dotilla* crabs, only a few centimetres long, behave in a most puzzling manner, for after spending some time in cleaning out their burrows, they proceed to carry small balls of mud to a distance of several inches and deposit them in circles around the burrows. Hans Hass has described how he and his wife Lotte passed many hours lying prone on the hot sand of a Great Nicobar beach, with cameras at the ready to film and photograph these crabs in the act of engineering their sand castles:

> As soon as the tide has left the sand bare the crabs appear on the surface and zealously begin to scrape up the top layer of sand, pass it through their complicated mouth-parts to extract the organic substances it contains, and dispose of the rest in the shape of little balls. When the ball reaches a certain size one of the claws hurls it away backwards. But the astonishing

Dotilla, *or soldier crab;*
another Dotillas' *uninhabited burrow in background*

thing is that these balls are not just tossed at random but in such a way that concentric castle walls are often formed.

Their film demonstrated that these embattlements, erected in a progressive series of arcs—as many as eight or nine in some instances—between burrow and feeding place, are not fortuitous, for it showed one *Dotilla* adding a number of pellets to a wall and then, while scuttling off to the next wall, untypically tossing a pellet into the fairway: whereupon it suddenly stopped, turned back, and pushed the pellet on to the nearest wall. Hass was unable to advance any good reason for this behaviour, though he speculated that the walls might serve as protective barriers if the crab was cut off from its hole—reef herons and even Caspian terns, which normally obtain their food by diving, prey on the *Dotillas*. But he subsequently noted that, on some Malayan beaches, only a few of the crabs formed a regular pattern with their pellets, the holes of others being surrounded by untidy, formless aggregations of pellets. A year or two prior to Hass' observations a German biologist, R. Altevogt, had also been much puzzled by the somewhat similar behaviour of *Dotillas* on an Indian beach. In their case, however, there were days when all the crabs were busily constructing "beehives" of pellets over their holes, while on other days none were doing so. Possibly the walls and beehives are a form of territorial demarcation, for Altevogt observed that there were occasions when some males would destroy the beehives of other males, and Hass perceived that if two crabs met in the course of building operations they would halt and brandish one red claw threateningly.

The scarlet and deep violet-blue rock crabs with lavender-coloured eyes—the "Sally Lightfoot" of the West Indies—

are perhaps in process of transition from a marine to a terrestrial existence. Klingel noted that on Inagua, where they ranged in size from a mere ½ inch to a massive 8 inches from claw to claw, their habitat was demarcated by the cliff tops on the one side and by the roaring surf on the other. They were not to be found anywhere else on the island. Within that restricted zone they obtained their food, mated, laid their eggs, and died. Nevertheless, while confined to a strip of rock within 20 feet of the edge of the sea, they would remain out of the water for hours at a time; but, like the hermit crabs, they were still tied to the sea by the need to dampen their gills at frequent intervals. By pressing their bodies close to the rock, so that the sea could not get underneath them and tear them off, and by gripping interstices firmly with their sharp claws, they were able to resist the tremendous impact of the surf. Moreover, since their stalked eyes enabled them to cover an arc of almost 360 degrees, they were never caught napping by a breaker. At the precise moment that one struck they flattened out until it was past, then raised themselves again and sidled to another position before the next wave arrived.

Nor have the orange land crabs, common inhabitants of many of the more arid islands of the Caribbean, entirely divorced themselves from the ocean. Klingel has given an account of a memorable night in Inagua:

> Up on the white beach, gleaming silver in the moonlight, were moving small shadowy forms. In long windrows they were gliding out of the dark bushes and creeping down to the surf. Faintly I could see the glisten of the salt water as it reached their bodies and drenched them with its coolness. For a second their forms showed half smothered in foam and then they disappeared. . . . The shadowy forms were land crabs. . . .

> They did not belong on the sea-shore, but . . . in those dry portions of the island where big cacti reared their heads above the soil . . . as far back as the borders of the great central lake where they rambled about the tangled thorn glades and the barren savannahs seeking the twigs and bits of green vegetation on which they feed. . . . No rain had fallen for weeks. . . . Back in the interior it had become very dry and the mud had split in jagged cracks. . . . Strange that they should be there on the beach. . . . Then I remembered. . . . Far back in the hinterland of the island, miles away, it had rained that day, a downpouring, drenching, tropical rain that filled the dry salinas to overflowing. . . . This rain was what the crabs had been waiting for, hidden deep in the cavities of their holes. And when the precious water came down, wetting their bodies . . . these crabs knew in some unexplainable way that their hour had come. . . . A sudden urge had taken hold of them, all at once. . . . The time had come to return to the sea. . . . In the moonlight I cornered one. . . . Beneath its body was a great purplish mass tucked under the apron. . . . I let her go and she dashed the remaining distance to the surf. . . . In some dark crevice the mother would shake off her burden of eggs. Just below the area of the surf . . . the eggs would hatch.

On Ifaluk Atoll the robber land-crabs go down to the sea to spawn on three nights after the full moon in the summer months of May, June and July. Are they stimulated to do so by the tidal waters rising and falling in their burrows or by fluctuations in the intensity of the moon's light when they are foraging abroad at night? Tidal rhythms influence the lives of all the creatures of the reefs, from the mollusks of the intertidal zone to the sea-birds nesting on the crowns of the cays. Fiddler crabs, which emerge from their mud burrows to scavenge for food during the period of low water, possess both tidal and solar timing mechanisms. Their bodies are dotted with hormone-controlled pigment cells, which at night are pale ivory, by day a dark brownish-gray; but they are paler coloured at night when the tide is high than when it is low, and darker when the tide is low by day than when it is high. This colour rhythm presumably serves, or once served, some camouflagic purpose of fundamental importance, to judge by its complexity; though one cannot begin to understand what purpose, for even a leg, shed during the course of a fight, will continue to exhibit without interruption the colour changes appropriate to the solar and tidal rhythms.

Those giant land crabs, the coconut or robber crabs, are widespread throughout the Pacific and Indian Oceans, except on very dry atolls with insufficient supplies of palm nuts and scavenging matter. These, too, are not entirely divorced from the sea. Even on the Galápagos Islands, where they inhabit the interior, they are reported to visit the sea at intervals in order to dampen their gills, and while little appears to be known of their breeding cycle, the thousands of eggs carried in the female's apron apparently hatch immediately she immerses her tail in the sea, when the larvae either remain in coastal waters for some time or are carried by the currents to other islands. Although robber crabs feed on a variety of nuts and fruits, including those of the pandanus palm, the shells of which are almost as tough as those of coconuts, fallen coconuts are their main source of food. Very large robber crabs may measure 2 feet in length and 6 inches across the carapace, span 3 feet between the tips of their first pair of walking legs, and weigh 5 or 6 pounds. Powerful enough to rip off the lids of strong biscuit tins, even though these are fastened down with wire, these crabs open coconuts by first tearing off with their pincer claws strips of the husk, fibre by fibre, from the end in which the three eye-holes are situated, and then hammering with one finger of the heavy claw in an eye-hole until an opening has been made through which it can extract the meat with its narrow pincers. Though capable of climbing to the top of a palm with the aid of its pincers and fourth pair of legs in order, according to natives of the Caroline Islands, to eat the delicious young leaves, there does not appear to be any record of them cutting down the nuts; but surprising quantities of the husk from fallen nuts are to be found lining their deep burrows under tree roots. If the appetite of a robber crab in captivity is any guide a single crab consumes 250 coconuts a year, the produce of three palms, under natural conditions; but no doubt the availability of other sources of food would reduce this potential consumption.

The robber crab has a much more significant claim to immortality in the record books than its consumption of coconuts. During the period 1946 to 1958 the U.S.A. conducted fifty-nine nuclear test explosions in western waters of the central Pacific. However, in November 1952 the first hydrogen device was detonated at Eniwetok Atoll on an islet about 1½ miles from the coral island of Engebi. Its explosion obliterated the islet, leaving nothing but a crater a mile wide in the reef, and rattled the foundations of Engebi, destroying the latter's vegetation with a fire-ball, and also submerging it beneath a wall of water. Engebi was also exposed to massive amounts of radioactivity, exceeding 3,000 roentgens during the first hour after the explosion. Now, in order to build up their shells, which are molted periodically, robber crabs require large amounts of calcium; but the chemical structure of calcium is very similar to radioactive Strontium-90. The crabs on Engebi which, as it happens, is deficient in calcium, must therefore have absorbed large amounts of strontium. Furthermore, if the habits of crabs in captivity reflect those of crabs in their natural habitat, they eat their molted shells piece by piece and would therefore continue to accumulate strontium in their systems. Yet, subsequent examination indicated that these Engebi crabs were apparently unaffected by radioactivity.

Even more significant, since the animals involved were mammals, were the reactions of Engebi's population of rats which, prior to the test, comprised two species, the Polynesian and the roof-rat or ordinary ship's black rat. For some weeks after the explosion the only rats to be found in the area were on an island 9 miles from Engebi, and these were so debilitated by radiation sickness that they were caught without traps. Yet the extraordinary fact is that a proportion of the Engebi population had in fact apparently survived the explosion, the fire-ball, the "tidal wave" and exposure to radiation—assuming, that is, that those found on Engebi three years later had not immigrated from neighbouring islands, the nearest of which was separated from Engebi by more than 300 yards of water flowing swiftly over the reef. Moreover, these survivors, or their progeny, exhibited no radiation abnormalities; nor indeed was there any abnormality in any of the island's fauna that could be assigned to radioactive causes, though detectable amounts of both strontium and Cesium-137 were present. By 1964, twelve years after the test, the colony was thriving to such a degree that there was some evidence of heavy population pressures on the available food resources. Moreover the nesting population of sea-birds was estimated not only to have recovered but to have increased, while although masses of fish had been killed by the explosion they too appeared to be back to normal in species and numbers.

If the field and laboratory research on the after-effects of the Eniwetok nuclear test has been precise, then this is proof of Nature's remarkable resilience and a ray of sunshine in the all-pervading ecological gloom.

6: Aldabra and Its Giant Tortoises

We can now consider some of the problems posed by oceanic groups of islands and atolls with notably specialized faunas, and in particular the Seychelles and the Galápagos. The ninety or so islands of the Seychelles scattered over half a million square miles of the Indian Ocean fall into three loose groups. The most northerly islands of the group, Mahé and Praslin, together with their associated islets and rocks, are unique among oceanic islands in being composed of immensely ancient crystalline granite knobs, 650 million years old perhaps, rising from a submarine plateau more than 10,000 miles in extent. With their red and brown, tree-clad mountains and lush verdure they have been compared by some travellers to such Polynesian paradises as Bora Bora and Moorea, though there is no geological affinity. Indeed there are no clues to the origin of these northern Seychelles. Although great deeps separate them from India and Africa, and from the other groups of islands, it is possible that they were once a part of a vast continent linking India with Madagascar, which also has a granite backbone, and southern Africa.

In the second group are the Amirante Isles, 150 miles south-west of Mahé. These are predominantly elevated sand-banks fringed with coral reefs and topped with guano. In the third group, though 700 miles south-west of the northern group, are Aldabra and its associates—Cosmoledo, Astov and Assumption—rising individually from depths of more than 12,000 feet. In contrast to the northern islands Aldabra is a relatively new atoll, perhaps only 50,000 years old, whose foundations are typically the remains of a coral reef based on a submerged volcano. It is also one of the world's largest atolls, with a lagoon 18½ miles long and 5 miles wide, most of which dries out at low water. But although technically a perfect atoll it has the external appearance of an island, for its land rim is perforated only by a few narrow channels, through which the sea floods into and out of the otherwise land-locked lagoon, and which split the atoll into four parts. Unlike most atolls, whose sandy crowns lie virtually flush with the sea, Aldabra has been uplifted, either by submarine volcanic eruption or by a fall in the level of the ocean, so that its coastline is encircled by 15-foot cliffs, though its land rim is slowly being eaten away both by the sea and by the lagoon. This exceptional elevation must have influenced Aldabra's ecology by enabling plants and animals to establish permanent communities without interference from the sea, and must certainly account for the existence of its extra-ordinary population of giant tortoises.

Uplifted atolls are uncommon, and Aldabra and Henderson Island (one of the Pitcairns in the south Pacific) are the only two whose flora and fauna have not been fundamentally changed and exploited by man and his domestic stock. Those of Aldabra owe their preservation mainly to a fortunate accident of geography, which has resulted in normal trading routes bypassing the island. Moreover, its dangerous shores have always been hostile to sailors, for Aldabra is protected by huge slabs of coral, which have been undercut by the sea and eroded into large tables supported by thin stalks, like gigantic black mushrooms; and whenever a stalk has been worn through its table has toppled into the sea to form part of a reef of jagged rocks capable of tearing the bottom out of any boat. Furthermore, the difficult terrain and lack of adequate soil and water has discouraged potential settlers. Thus, with the exception of an occasional ship calling to take on food supplies in the form of turtles and tortoises, Aldabra was left in peace during the seventeenth, eighteenth and nineteenth centuries when the more accessible and commercially attractive islands in this sector of the Indian Ocean, such as the remainder of the Seychelles and the Mascarene group of Mauritius, Réunion and Rodriguez, were stripped of their indigenous forests and planted with such commercial crops as sugar-cane and coconut palms. This destruction of their natural habitat, together with the introduction of such harmful animals as macaque monkeys, dogs, rats, pigs and goats, inevitably resulted in the extermination of much of the native wildlife during the seventeenth and eighteenth centuries, including such vulnerable flightless birds as the dodos of Mauritius and Réunion and the solitaires of the latter island and Rodriguez, together with more than thirty other mainly flightless species of birds, leaving a rail (*Dryolimnas cuvieri*) as the last survivor of the Indian Ocean's flightless curiosities.

Flightless white-throated rail in aggression display

More recently, the importation to the Seychelles of barn owls from East Africa, in an attempt to control the numbers of palm rats and tenrecs in the coconut plantations, has proved to be yet another catastrophic example of introducing an animal to a new environment without first weighing all the potential effects of such an introduction, however improbable these might seem. The clearing of the indigenous flora and the drainage of swamps on the Seychelles, with the consequent loss of reed beds and mangroves, had resulted not only in a great increase in the rat population in the plantations, but had also deprived the islands' birds of much of their natural habitat. As it happens, the initial introduction of barn owls in 1949 was unsuccessful, apparently because they were accustomed to hunting on the open African grasslands and were unable to secure the Seychelles rats in the thick cover of the plantations. These owls all died; but subsequent imports in 1951 and 1952, after a difficult initial period when they subsisted on cockroaches, eventually established themselves and, having found a new prey, in addition to rodents, multiplied and spread from island to island with extraordinary rapidity. Their new prey were sea-birds, and in particular the fairy terns which, unlike other terns, lay their single eggs in the most precarious positions from ground-level upwards, whether it be in a slight hollow in the bark on the bough of a tree or in a crotch where two branches meet, on a projection of rock, or on a fallen palm frond. Although the precarious nature of these nest sites is offset to some extent by extremely close sitting on the part of the parents, both eggs and chicks were already being heavily preyed on before the arrival of the owls, by crabs, skinks and another Seychelles alien, the cattle egret. Though a natural immigrant in 1915, the egrets, like the owls, exploited the food resources of their new environment during the fairy terns' breeding season by changing from their normal insect food to a diet of eggs from unattended terns' nests. Being snow-white and on the wing among the casuarinas on all but the darkest nights, the fairy terns have proved easy prey for the barn owls and have been exterminated on some of the islands.

Another example of the havoc wrought among indigenous fauna by an importation of an alien species is that of the South American giant toad, which was introduced to Australia in the 1930s as a means of controlling the destructive sugarcane beetle. Subsequently it dispersed more than

1,000 miles down the Barrier Reef coast, not only displacing the native frogs but, being poisonous, resulting in heavy losses among the waterfowl and such snakes as adders, pythons, and tree snakes preying on this unfamiliar creature.

However, we are particularly concerned here with the life and death of the giant tortoises. From the seventeenth century they provided the crews of European ships with a staple food supply. At the end of that century they were so numerous that they were reported in carpets of several thousand on Rodriguez, and 30,000 were shipped from that island to Mauritius in 1759–60 while, in 1826, 2,400 were shipped from the Seychelles and 800 from Providence. By 1850 they had been almost exterminated on some thirty islands in the Indian Ocean from Madagascar to the Chagos Archipelago, and today they are left with only two habitats in the world —Aldabra and the Galápagos. A French captain, Lazarre Picault, who was despatched from Mauritius in 1742 or 1756 to explore the possibility of occupying the Seychelles, has left a curious report of crocodiles killing tortoises in the mountain gorges of Mahé. This does not sound a particularly credible story. Nevertheless, it was an unlikely one to invent, and we know that crocodiles of the Nile species were in fact common in the brackish coastal marshes and in some of the inland swamps and streams on various of the Seychelles until early in the nineteenth century. These presumably reached the islands by sea, for crocodiles are strong swimmers, propelling themselves mainly with their powerful tails, though sometimes making use of their partially webbed feet when paddling slowly; and the estuarine crocodiles of the Indian and Pacific oceans, which have been sighted 40 miles out to sea, must be capable of oceanic crossings, for they have colonized the Cocos-Keeling Islands 560 miles from the nearest island of Sumatra and have also been reported in Fiji and the New Hebrides.

But how is one to explain the presence of giant tortoises on the far-flung islands and atolls of the Indian Ocean? Since Aldabra is a volcanic atoll it can never have been joined to Madagascar, 250 miles distant. Tortoises are known to have existed at one time on that island which, after its separation from Africa some 20 million years ago, became a repository of relict animals, notably of various species of lemurs and that gigantic bird the *Aepyornis* which may have survived until the middle of the seventeenth century. Though less than 10 feet tall and therefore shorter than the Moa of New Zealand, *Aepyornis* was considerably heavier, reaching a weight of possibly 1,000 pounds, and producing eggs more than 12 inches long and twice that girth. Rodriguez is nearly 1,000 miles from Madagascar and considerably more than 2,000 miles from India.

Giant tortoises are efficient swimmers. If one is placed in the sea it floats buoyantly, barely drawing water, with plastron resting lightly on the surface and head and neck sticking up like a periscope. Even its legs are hardly covered as it attempts to propel itself with the extremities of its bowed forelegs. William Travis, who visited Aldabra while collecting green sea-snails for their mother-of-pearl, found that when he immersed a tortoise forcibly under a coral overhang for five minutes it remained motionless, not even blinking an eye, while a tiny bead at each nostril indicated that it was holding its breath. Theoretically therefore, giant tortoises might be capable of long ocean voyages. But although they amble along the beaches of Aldabra and into the shallow lagoon, in which they cruise around freely while submerging their long necks in order to feed on the trailing bottom weed, and although there are old local reports of them drifting in the currents from one island to another of the Seychelles, there does not appear to be any definite record of a tortoise actually being seen to enter the sea voluntarily.

Theoretically again, though improbably, numbers of tortoises might be afloat in the Indian Ocean at one time, after having been carried out to sea by floods or sucked out by a tsunami; and since they can survive without food or water for several months, they should be able to remain afloat in the warm tropical waters for long periods. However, it is not even certain that they could remain afloat for any length of time, for a Galápagos tortoise, which William Beebe put over the ship's side for some hours, subsequently died from congestion in the lungs and intestines, as a result, he surmised, of the amount of salt water it had swallowed. But assuming all these improbables to be possible, and the currents in the Indian Ocean are all against a favourable dispersal of tortoises, there still remains the problem of actually establishing a new colony, which would have to be founded either by a female bearing fertilized eggs, by a male and a female landing on an island together, or by one or the other waiting several years for a mate to be cast ashore providentially.

All these possibilities seem wildly improbable, and one is more impressed by the fact that all the former tortoise islands in the Indian Ocean lie on or near the various submarine ridges that, with westward branches, extend from India to the Antarctic, and which at one time might have been elevated sufficiently to provide a progressive series of islands and archipelagoes.

Aldabra's peripheral uplifted reefs have been intricately dissected by salt water into innumerable jagged pinnacles and hard black razor-sharp ridges, crevasses and solution-potholes 12 feet deep. This is the infamous *champignon,* which cuts through the sole of a rubber boot in a day. It is interspersed with the *platin:* slab-like flagstones which ring musically underfoot and which are weathered coral detritus deposited on the landward side of the reefs, and impregnated by the guano of millions of sea-birds. When one adds that the *champignon* is covered with dense thickets of pemphis thorn through which a path can only be hacked with the aid of machetes, and that shade temperatures exceed 100 degrees F, then it becomes clear why the Creole labourers on the plantations have not exploited the birds or tortoises of the interior; why their goats, cats and rats have apparently been responsible for exceptionally little damage to the flora and fauna; and why the interior of Aldabra remained virtually unexplored until the mid-1960s. Indeed it was authoritatively stated as recently as 1959 that, apart from the Galápagos, giant tortoises were to be found only on islands off south-east Africa, though no longer in the wild state; and seven years later, that they were almost extinct, surviving, if at all, only in captivity. Yet, when the ubiquitous Francis D. Ommanney had visited Aldabra in 1948 he found every part of the island except the *champignon* densely populated with tortoises, with dozens under every bush and tree; and where the scrub gave way to a rolling plain of almost English-looking turf they were strewn around in hundreds, like smooth round multi-sized boulders, on the loose sand beneath the stands of gray-boled mapon trees, whose broad soft leaves afforded shade.

However, the 1959 report that the tortoises were extinct in the wild state on Aldabra, because the goats had denuded the vegetation, led to an expedition from the University of Bristol in 1964–5, only the second expedition to visit Aldabra since Darwin's day. Its members' discovery that the tortoises were very much alive, and their estimate that there were in fact about 30,000 breeding on Aldabra, in densities of almost 5,000 to the square mile, must have been a source of some pleasant embarrassment in scientific circles. Moreover, after living on Aldabra for twelve months, Tony Beamish raised the estimated population to between 60,000 and 100,000! Clearly, these figures represent no more than informed guesses; but at any rate they confirm that, happily, there are still large numbers of tortoises on Aldabra. Since the Galápagos population is certainly not less than 3,000 and possibly exceeds 10,000—estimates vary from one year to the next—Aldabra thus holds the bulk of the world's existing stock of giant tortoises, protected hitherto by its hostile terrain, but still threatened by the possibility of the atoll being developed as an air-base or tourist Mecca.

There is some doubt about the all-important status of goats on Aldabra. Although Ommanney had seen a herd of more than a hundred in 1948, Beamish apparently did not see any and only came across their droppings on one occasion. It was his opinion that they could not survive on Aldabra in any numbers because, during the long dry season, any living vegetation is virtually restricted to the vicinity of the few permanent freshwater pools. However, knowing a goat's ability to eke out a living in the most barren country, this argument is not convincing, and in 1956 William Travis, who described an entire cliff edge as being a solid mass of goats, was impressed by their immense size; but according to him, they are more or less marooned along the south-east segment of the atoll, where there are no tortoises. This is barren *champignon* country with little vegetation except screw-pine and a tangled mass of pemphis in which the goats usually live, though when dense clouds of mosquitoes hatch out during the rainy season, they are obliged to move out to the rocks and gulleys of the foreshore.

The tortoises themselves can no doubt survive many weeks without food and water, and during the dry season can eat and apparently digest any matter from dead leaves and scattered tufts of withered grasses and scrub and the remains of crabs, goats and other tortoises, which perhaps supply them with essential minerals and extra proteins, to such impedimenta as paper, tinfoil, cigarettes, vitamin pills, and even paint. They are also reported to be attracted like turtles, to any red object. During the rainy season they feed, like their small continental *Testudo* relatives, on the fresh young grass which grows as a thick turf on the landward side of the sand dunes, and in patches where the scrub is not too thick on the *platin*; on water-weeds in the shallow pools frequented by flamingoes; and, in coastal regions, on the leaves of shrubs, especially the laurel-like *Scaevola*. But they are selective in their choice of these vegetable foods, and Beamish observed that when there were still adequate supplies of greenery towards the end of the rainy season, several tortoises might be seen converging on a single small plant on a stony plain. He also saw evidence of what he considered to be a desperate search for food, with tortoises rearing up on one another's backs, and extending their necks a foot or more to reach green leaves, while most trees and large bushes were marked with a crop-line as if they had been grazed by a herd of diminutive cows. However, this again was probably a matter of preference rather than of necessity, because other plants were apparently left untouched during this intensive browsing on leaves. Nevertheless, it is worth asking the question—has the Aldabra population of giant tortoises reached saturation point? Are they in danger of eating themselves out of house and home?

Beamish stated that he could find very few seedling trees, and how is a woodland to survive if not able to regenerate? This is a major factor to be taken into account when considering the conservation of the tortoises, for trees not only provide them with food but also with the no less important shade. Any tortoise caught out on the beach in the heat of the day soon dies if unable to reach shade. So, after feeding in the early morning, the tortoises retreat into the shade of a grove of trees, some to half bury themselves in the mud of a pool, when the air temperature reaches the low-80-degrees F. In the evening they come out again when the temperature is approximately the same and feed until nightfall, when they stretch out their yard-long necks along the ground since there are no predators to fear, close their eyes, and sleep wherever they may happen to be with all four legs straddled. Man himself, walking through a group of tortoises, provokes no reaction. Travis has described how when he returned, after experimenting with his tortoise in the sea, its fellows were all in the exact positions in which he had left them earlier:

> The biggest one still had half a blade of grass protruding from its mouth. At what point in the past it had ceased eating I did not know, nor when in the future it might continue. I had an eerie feeling that this sunlit clearing was somewhere out of time. . . . A heavy silence permeated the air and not even the weird lost cries of strange birds away in the thickets of pemphis seemed real. A white Ibis stalked among the sleeping giants, its grotesque struttings brought into sharp relief by their immobility. No breeze stirred the grasses here. A vivid green lizard crouched motionless on a flat stone by my feet; its hard agate eyes, extended tongue of jet and metallic sheen of skin made it not a thing of flesh and blood, but a lifeless exotic bauble.

On cool cloudy days or after rain the tortoises may be abroad and feeding in mid-afternoon, as may those occupying shady groves with freshwater pools, when those elsewhere are sleeping. Their greatest concentrations are located in the most developed mixed woodland, especially, as in the case of most of the island's birds, in the groves of banyans and hard evergreen takamaka where the freshwater pools are not contaminated with the salt spray of tidal water. But large numbers are also to be found in areas where the pools are brackish, or even where there is no water at all, on barren wasteland and sandhills, providing that, again, mixed woodland or beach vegetation is at hand to afford the essential shade.

But, to return to the question of feed resources, how could the tiny island of Rodriguez, considerably smaller than Aldabra, support a tortoise population from which 30,000 could be shipped in one year, particularly in view of the fact that these tortoises had no natural predators once they had passed the critical stage of adolescence, and lived to a very great age? Little is known about their breeding habits on Aldabra, though the females perhaps lay their eggs in pockets of guano, or in the sand dunes; but during their first five years, until large enough to be more or less immune from attack by natural predators, the young tortoises are apparently subject to a massive mortality, and it is presumed that they are preyed on by such birds as pied crows, the sacred ibis and frigate-birds (which decimate young turtles in other parts of the world), but especially by robber crabs and no doubt by rats and the few feral cats and dogs on the island. Those that survive do so by their surprising agility and ability to conceal themselves in cover very quickly, and since there are no records of giant tortoises suffering from any diseases or parisitical infestations the survivors can look forward to long and peaceful lives, though a few become casualties. Some are trapped in the steep-sided potholes, filled with vegetation or overgrown with it, on the borders of the spiky *champignon*. Three skeletons and five living tortoises have been found in one such pit, for they can survive in such a place for months before finally succumbing to dehydration or starvation, whereas those that slip into rocky crevices and become immovably wedged, die quickly in the heat of the sun, and are soon reduced to empty shells by crabs, crows and other tortoises. Although it is possible to establish the ages of a proportion of the young tortoises for a number of years by the annual growth rings on their carapaces, these subsequently become too worn to be deciphered accurately. However, a giant tortoise is believed to mature at the age of twenty or twenty-five years, when some 2½ feet long, and thereafter to grow very slowly at a rate of about half an inch a year. Only adult males can be reliably sexed, and today the largest of these on Aldabra reach maximum weights of 400 pounds and exceed 40 inches in length, though there are earlier records of tortoises twice this weight and 4 feet long, with the crown of their domed shells reaching a height of 3½ feet above the ground, and still larger ones may have been slaughtered in the past on other islands in the Indian Ocean. On the Galápagos there are old reports of tortoises exceeding 5½ feet in length and 4½ feet across the carapace, and William Beebe noted that ages of 400 to 500 years were claimed for these giants. Whatever may be the truth of such claims, there does not seem to be any doubt that these tortoises are capable of becoming centenarians. A tortoise on Mahé was reputed to be more than 152 years old in 1970; a Tonga tortoise, reputed to have been seen by Captain Cook, and to have survived the loss of one eye, the fracture of its shell by a kick from a horse, and two forest fires, would have been at least 176 years old in 1950; and of two tortoises from Aldabra, reputed to be adult at the time of Napoleon's exile on St. Helena, one fell over a precipice in 1938 at the age of 123, while the other was still alive in 1948 when 133 years old. A fifth, specifically mentioned in the 1810 treaty under which France ceded Mauritius to Great Britain, was reputed at that time to have already lived on that island for at least seventy years, and since it was still alive well into the twentieth century would therefore have reached an age of at least 180 years; while a sixth Mauritius tortoise, originating on Aldabra and shipped to England in 1897, is reputed to have been between 150 and 200 years old at the time of its death, when it weighed 560 pounds. Finally, a seventh from the Galápagos, shipped to Honululu before 1850 and subsequently to England in 1915, died two years later at a minimum age of sixty-seven.

7: The Giant Tortoises of the Galápagos

Let us now travel to that other home of the giant tortoise, the Galápagos Islands, 660 miles west of Ecuador and some 3,500 miles, by dead reckoning, from the nearest of the Pacific islands, the Marquesas. There is no conclusive evidence as to whether the Galápagos are volcanic in origin or were formerly joined to the mainland of South America; but since their basaltic lava is pitted with some 2,000 volcanic craters and studded with a myriad mounds of fragmentary rock, one must assume that they were created by volcanic eruptions. Moreover, what evidence there is suggests that they have been elevated from the sea rather than depressed from a continent, for as the geologist Lawrence J. Chubb has pointed out, there lies beneath the eastern section of the central Pacific the immense Albatross Plateau in deeps approaching 12,000 feet; and while no islands rise from the central areas of this plateau, archipelagoes at either end of it—the Marquesas at the western end and the Galápagos at the eastern—appear to have been built up on a set of intersecting fissures. Moreover, on or near the southern margins of the plateau are several true volcanic islands, including Maugerera, Pitcairn, Easter Island, Sala-y-gomez and Juan Fernandez.

On the other hand, for the Galápagos to have been a part of South America would have involved the submergence of the intervening land, or a rise in the level of the sea, by some 12,000 feet. There are no indications of such a cataclysm. Nor does the flora and fauna of the Galápagos show any evidence of being continental in origin. The most notable feature of the flora is the scarcity of tropical American plants and the predominance of *Compositae*, whose record in other parts of the world is of colonization by long-distance oceanic dispersal. Moreover, as Bryan Nelson has pointed out in his study of sea-birds on the Galápagos, the numbers of invertebrates are extremely restricted, whereas if a land-link had existed within the last 2 or 3 million years, many of the invertebrates, so ubiquitous in South America, would surely now be found on the Galápagos. Many primitive leaf-mould beetles are, for example, absent from the latter though common in similar habitats in Ecuador. The varieties of mollusks are also limited, as they typically are on oceanic islands, and one that does occur is representative of a group widespread throughout Polynesia, but not found anywhere in South America. Furthermore, the Galápagos have no amphibia, which are also typically absent from most oceanic islands, because neither they nor their spawn can withstand immersion in salt water, though it is true that three species have found a means to reach the Fiji Islands and a fourth the Sandwich Islands. Again, despite the varied habitats of mangrove swamp, cactus semi-desert, humid forest and open uplands on the Galápagos, there are only three native mammals—a paucity to be expected on an oceanic island—the rice- or cane-rat and two species of bats. The latter, with their power of flight, are natural immigrants—there are five species on Aldabra—and it is significant that those found on the Galápagos are migratory and insectivorous species that do not enter into true hibernation; for one is the hoary bat (*Lasiurus cinereus*) which summers in the temperate regions of North America and migrates southwards, and often coastwise, in the autumn before the temperature becomes too low for insects to be on the wing; the other a close relative to the red bat (*Lasiurus borealis*) which migrates in the fall to the coast of California where insects are flying throughout the winter. Another *Lasiurus* has colonized Hawaii, the only bat to do so.

Nor do geologists favour the hypothesis that the submarine ridges linking the Galápagos to Ecuador and to Cocos Island, 500 miles north-east, were formerly land. This possibility cannot, however, be ruled out. In 1963 the bones of a new species of rat-like rodent, related to a group previously known to inhabit only a few islands in the West Indies, were discovered in a cave on the Galápagos island of Indefatigable (or Santa Cruz); while the presence of as many as five kinds of reptiles—giant tortoise, marine and land iguanas, a snake, a lizard and a gecko—is untypical of an oceanic island, but would be compatible with the existence of land-bridges, complete or interrupted, such as might be provided by uplifted submarine ridges. It would have been feasible for pioneer iguanas to have migrated from island to island along disconnected land-bridges, or indeed perhaps to have crossed the 600 miles or more from the South American mainland under their own power, though, as we shall see, they do not now appear to venture more than a few hundred yards from their island homes. We have already discussed the improbability of giant tortoises surviving trans-oceanic voyages, though it is conceivable that small *Testudos* might have made fortuitous sea-crossings on driftwood "rafts." But the latter, and still more so lizards, geckos and rice-rats, could have been transported to the Galápagos on the balsawood rafts of the Inca and pre-Inca peoples who repeatedly visited the islands.

The point is that the Galápagos are not the repository of very primitive reptiles such as dinosaurs that have become extinct; for while it is true that, in the virtual absence of competitive mammals, reptiles are the dominant terrestrial fauna on the islands—as was once the case over the remainder of Earth—these are modern types of reptiles. The Galápagos iguanas are related to the contemporary iguanas of tropical America, and the giant tortoises to the common *Testudos* which still inhabit both the West Indies and South America. The Galápagos have been in existence for not less than 2 million years, and are probably considerably older than this. Fossil giant tortoises, perhaps 60 million years old, have been located in Cuba and apparently Patagonia; but on the assumption that the Galápagos are true oceanic islands, and that therefore tortoises are unlikely to have reached them in their present gigantic form, it would not have been impossible in this space of time for this gigantism to have evolved from a race of small *Testudos* in an environment in which they were vulnerable to predators only when small, so that it was an advantage to grow larger. The period of isolation in a new environment required to develop physical differences has, arguably, been vastly overestimated.

Although the Galápagos' fourteen sizeable islands and fifty-odd islets and rocks are scattered over 23,000 square miles of ocean, their total land mass is only about one-sixth of this, for the largest island, Albemarle (Isabela), which includes more than half the land area, is only 75 miles long and from 25 to 45 miles wide. All have apparently the same geological structure: yet are obviously of different ages. The small outlying island of Tower (Genovesa), for example, 90 miles north-east of the central group of larger islands, exhibits razor-sharp lava fields: whereas those on another outlier, Hood (Espanola), a similar distance south-east of the central group, have been weathered to smooth red slabs and boulders; and those on Charles (Santa Maria), another outlier between Hood and the main group, have been still more severely weathered and broken down into a rich black soil. The Galápagos must therefore either have been uplifted volcanically at different times or separated by subsidence. They are still volcanically active. On Albemarle five volcanoes, evenly spaced, extend almost the whole length of the island and their lava-flows curve down from the breached craters like black glaciers. When William Beebe's ship the *Arcturus* was lying off Albemarle at dusk on April 13, 1923, he noted in his log that the molten lava was creeping down the slope in true Pompeiian fashion. "The bivouac fires of a tremendous army seemed to be scattered over ten or twelve miles of country, and as the hours passed the whole black incline became daubed with slowly-writhing scarlet streams creeping towards the sea. The sun set almost directly behind the ridge, and the changing of scarlet sunset into rose and scarlet of cloud and lava was marvellous."

Two months later the lava had worked its way down to the steep cliffs of the shore, with nine great cascades of molten rock gushing from the face of the black cliffs, and dropping straight into the sea.

Hoary bat

Immense columns of steam were blown by the strong wind across the land, so that the cataracts were not obscured and we could watch huge pieces of the cliffs crumble under the pressure from behind, crashing outward to release fresh torrents of red-hot lava, that spouted like water from a culvert. Now and then submarine explosions from the too-rapidly-cooled lava threw great lumps of glowing rocks above the breakers, that hissed and turned to steam as they dashed against the scorching shore.

The sea was choppy with tossing white-caps; along the coast the water was livid light green, where it was heated by the lava; a line distinct as though painted on a floor, marked the beginning of the deep blue, normally cool ocean. So sharp was the demarcation that when the *Arcturus* was within a quarter of a mile of land . . . and lying distinctly across this line, her bow was in the green water at a temperature of 99, and her stern was in blue water which registered 78. As molten lava reached 3,000, the ocean under the cliffs was literally boiling. A sea-lion flung itself in agony from the scalding immersion, five times leaping all clear, and then seen no more. Shearwaters and frigatebirds stooped through the vapor to snatch at fish floating in this gigantic cauldron, and we saw dead petrels and shearwaters that had ventured once too often to this tempting feast.

Two of Albemarle's volcanoes continue to be active, with a violent eruption as lately as 1959, and there have also been recent eruptions on both James (San Salvador) and Narborough (Fernandina), culminating, after eight years' quiescence, in a multi-megaton blast on the latter island on June 11, 1968, following a relatively minor explosion on May 25.

The volcanic and arid nature of the Galápagos, particularly their coastal regions, has impressed all visitors. The gnarled trees, their twisted and stunted branches bleached by salt spray; the predominance of spiny and thorny shrubs, and of those with thick fleshy leaves or stems; the huge grotesque cacti, oval pads raised aloft, stiff and stark against black lava boulders have astonished them. They have been astounded too by the immense sheets and discs of clinker, resembling misshapen manhole-covers, balanced on edge or thrown together as the last upheaval of earthquake left them, on cratered Indefatigable (Santa Cruz), third largest island; and above all by Narborough's great hulk of clinkers smouldering in the banked fires of its towering volcanic cone.

Beebe was astonished at the apparent greenness of the arid coastal regions, where rain falls irregularly from January to April, until he discovered that in fact only the terminal twigs and branches of bushes and trees were in leaf.

> All a plant's or bush's energy shoots forth here. Parting the green skin, one saw at once the bare bones underneath—the wooded stems which were whitened in the months of rainless sunshine . . . sheltering deep beneath the armour basks the spark of life, which another rainy season, perhaps seven or eight months away, would arouse . . . the guard against the heat and dryness of all the other months had to be so complete, so hermetically perfect that it could not be broken even by the softening rains. Only the tips of the branches and the twigs were susceptible.

But very different conditions prevail in the interiors of most of the larger islands, for the green jungles of their middle zone and the fernery and lush grass of their highlands are cordoned off by the waterless lava belt of thorn and cactus in their desert lowlands. In the highlands lies another world, which has been compared to that of the Andes at an altitude of 10,000 feet—a cool world of rain mists all the year round, where the sky is nearly always obscured by the thick cloud of the *guara* mist created by the cool Humboldt Current, enveloping the hills and swirling through the high valleys. Long tresses of the pale green filamentous lichen, *Usnera plicata*, hang from the boughs of the trees; vines and lianas reach to the ground. Reddish beams of sunlight filter through the lichenous curtains to splash on the ferns. Yellow warblers and vermilion flycatchers, hawking insects over pools in hollows of the cratered hills, flash brightly through the filtered light. On a fine day in the uplands it is difficult to believe that one is on the Galápagos, so lush, in contrast to the deserts of the coastal zone, is the verdure of the forested hills encircling the pale-blue waters of the crater lakes.

These, then, are the contrasting environments in which the Galápagos' unique fauna and avifauna have developed their numerous island races, for the puzzling fact is that on many of the islands some of the birds or reptiles or plants may differ slightly from those of the same species on other islands, indicating that they must have been isolated on their respective islands for considerable periods, despite the fact that the central group of larger islands all lie within 25 miles of each other, and that none of the outliers is more than 100 miles from the central group. So far as the reptiles are concerned, this isolation may perhaps be attributed to the exceptionally strong currents which flow in the very deep waters around and between the islands, and which were responsible for the Galápagos being known at one time as Las Islas Encantadas, the Enchanted Isles, because the set of the currents persistently prevented ships from fetching up to the islands or even making a landfall. Island differences in the flora are perhaps explained by the fact that the Galápagos are a centre of phenomenal calm with very little spore- or seed-bearing wind, though these conditions cannot account for the absence of inter-island migration by the passerine birds, among which this island sub-speciation is particularly prevalent. One can readily accept the probability that the strong currents between the islands would have prevented the dispersal of the giant tortoises from one island to another, except by the rarest accident, though, if this has been the case, how is one to account for their presence on so many isolated volcanic islands uplifted at various times? One is bound to admit that, improbable though it may seem, geologically, this island sub-speciation in the fauna and flora is indicative of communities being isolated by partial submergence of an originally united Galápagos.

The splitting of a species into a number of sub-species has often been based on physical distinctions too slight to warrant any such splitting, and it is arguable whether there is really more than one race of giant tortoise, the *galápago*. However, some authorities consider that there are 6 island races; others that there are 11 or 12 distinct sub-species, including 5 isolated on each of Albemarle's volcanoes, and that 2 or 3 more have died since Darwin's time, when he landed on 4 of the islands in September and October 1835, and when tortoises were numerous on at least 10 of them. The most obvious physical distinction lies in the shape of the carapace, which conforms to two main types. The young giant tortoise's shell is as flat as the ordinary *Testudo*'s, but after about two years it begins to assume either a domed shape or a saddle-shape, and this difference appears to be associated with different habitats. Tortoises with the most pronounced domes (and relatively short necks) live, for example, on Indefatigable, Chatham (San Christobal), the southern part of Albemarle, and formerly lived on Charles. The highland regions of these islands are the most humid, and support the lushest vegetation, in the Galápagos; providing rich grazing and fresh water for at least part of the year. Grass is the tortoises' favourite food; but the acid and astringent leaves of the guava are also eaten, as is the lichen *Usnera plicata*, together with the leaves of various shrubs such as *Scalesia*, though not the abundant salt-bush *Cryptocarpus*. The *Scalesia* tree-sunflower, one of those long-distance migrants of the *Compositae* family, dominates much of the lower forest on the Galápagos, clustering in dense groves on the naked lava at between 600 and 1,800 feet.

By contrast, those tortoises with the most pronounced saddle-shaped carapaces (and long necks) inhabit Duncan (Pinzon), the northern volcano of Albemarle, and Abingdon (Pinta) which is studded with craters, one of which has recently been active. These are the least humid and most arid regions of the Galápagos, with no permanent grazing of grass or herbs. However, the effects of the absence of lush vegetation in these regions and in the prevailing arid coastal areas of other islands must not be overstressed, for in such habitats the tortoises feed mainly on cactus pads which, though 80 per cent water, have a moderately high nutritive content of sugar and starch. Rollo Beck, who lived for many months on the Galápagos in 1902 and 1905, was astonished by the soft-tongued tortoises' ability to eat the sharp-spined cactus pads, "but that they do so, and greatly relish them, is proven on several islands by the way the cactus leaves and blossoms disappear from under the trees. . . . On Duncan Island . . . the trails from one cactus tree to another were as plain and as well worn as cattle trails on a well-stocked ranch."

The tortoises of the bad-lands can also browse on the sparse foliage of such shrubs as croton, with its strong-smelling leaves, and the manzaria fruits; and, like those on Aldabra in the Seychelles, they eat the rotting carcasses and dried skins of the goats. It is said that, even when feeding on low plants, these saddle-back tortoises still reach up for the topmost leaves, the implication being that the long neck associated with the saddle-back, and also the upward flare of the fore end of the carapace permitting the neck to be elevated vertically, are the end-product of natural selection working on the survival advantages of being able to reach up to cacti and other shrubs. But, as we shall see, a tortoise is not necessarily permanently restricted to its particular humid or arid habitats.

On observing these giant tortoises of the Galápagos and Aldabra, one's first question must be, why have they achieved this gigantic stature? "Heavy as chests of plate, with vast shells medallioned and orbed like shields and dented and blistered like shields that have breasted a battle, shaggy, too, here and there, with dark-brown moss," in Herman Melville's words. There is no evidence that their gigantism was originally evolved as a protective measure against some large predator, but it is reasonable to assume that it could only have been achieved in an ideal environment, and that it is associated with good feeding and longevity. With no large predators to cut short their life-cycle, once they have got safely through the critical early years, the tortoises can grow and achieve sexual maturity at leisure. In their slow progression through life they remind one of the sloth; and just as the sloth has algae growing in its dense fur, so the male tortoise has lichen growing on a small crescent-shaped area at the hind end of its carapace, the only part of the shell not periodically immersed in water or scraped as the tortoise forces its way through the undergrowth, though no lichen can establish itself on the female's shell because this is scoured by the male's plastron during the act of mating.

Beck observed that the largest, and therefore the oldest tortoises, inhabited the richly vegetated southern section of Albemarle. There, the gradual slopes of the first 1,000 feet above sea-level are composed of rough lava, in the cracks of which trees and shrubs are able to take root and grow during the three or four months of the rainy season. In the course of the next 1,500 feet the vegetation is more abundant, the trees are heavily parasited with ferns and orchids, and vines and bushes make travelling difficult. At that height the forest ceases, and long rank grass and brake-ferns form the principal growth, with a heavy fog hanging almost continuously over the volcano throughout the summer. In these two zones, according to Beck, the majority of the tortoises spend their time from May until January. Observing that, whereas the shells of the smaller tortoises living near the base of the volcano were smooth and unmarked, those of the large tortoises near the summit were, almost without exception, irregularly pitted and scarred, he speculated as to the possibility of this marking having been caused by showers of lava many years earlier, since there was no evidence of recent volcanic activity.

Residents on Albemarle told Beck that the tortoises migrated to the summit of the volcano during the summer—their long sharp claws enable them to climb steep, rough lava slopes—and he noted that their trails extended for miles up and down and around the mountainside, from one grassy flat or rocky basin holding water to another. From centuries of constant pounding by their elephantine feet, as large as a man's clenched fist, the rocks had been worn so smooth that it was almost impossible to walk over them when they were wet after rain, while on the low sandstone ridges the trails were 2 feet deep and tortoise-wide. During his ascent of this southern volcano Beck counted some seventy-five tortoises, less than half-a-dozen of which he adjudged to be females, in the course of the 3-mile climb to the upper edge of the forest. The majority were in open glades and sunny parks, while in a little valley near the summit some were considerably larger than any seen at lower levels. Even at that height there were scattered trees and plenty of bushes to provide shade, together

Giant tortoise of the Galápagos

with an abundance of grass being grazed by large numbers of cattle, and several deep ponds.

More recently, in 1964, Eric Shipton, better known as a Himalayan mountaineer, found tortoises at an altitude of 3,500 feet "all along" the crater rim, 15 miles in circumference, of Albemarle's Volcan Alcedo, and also counted seventy-five in two days, only four of which were less than 18 inches long. Others were actually 1,000 feet down at the bottom of the crater, whose floor was a flat expanse of lava, two-thirds forested, with a lake whose boiling waters erupted every few seconds with geysers shooting jets of water 80 feet into the air. It has been suggested that there may be as many as 4,000 tortoises on Volcan Alcedo.

In addition to good feeding, ample supplies of water may also be essential if tortoises are to achieve their maximum growth. Those on the driest islands, such as Hood and Duncan, never apparently attained to great size. Although a fair amount of rain falls on the Galápagos highlands at all seasons, the volcanic rock is so porous that rainwater disappears immediately. Pools are therefore rare and streams non-existent. Indeed, the only island with a permanent supply of fresh water is Chatham, where water from rain and heavy mist is retained in one of the high, broken-down craters; though, if Melville is to be believed, there was good water on Barrington (Santa Fé), and possibly also on Charles, in the first half of the nineteenth century. On Indefatigable the largest dome-shelled tortoises inhabit the pond and grass country of the highlands. On Albemarle, Beck observed that one of the objective points of their trails was a rock basin in which water collected during the rains. And Darwin, noting that springs were only to be found at considerable altitudes on Chatham, observed that tortoises from all the lower regions travelled long distances, plodding perhaps 4 miles a day with their slow, regular heavy gait along numerous broad, well-beaten trails, in order to drink at these pools, plunging their heads into the water and gulping great mouthfuls at the rate of about ten a minute. Those living permanently in ponded areas pass much of their time in the water or, when the pools are dry, in mud wallows. When sheltering under a tree during a heavy downpour one afternoon, Beck was surprised to see a large tortoise come slowly down the hill through the wet grass, walk into a rapidly forming pool and, after taking a long drink, lie down in the pool. An American camera man, Jack Couffer, has described an encounter with a large tortoise, whose carapace was protruding like a black island from the depths of the mud in the middle of a watering place:

> A tiny head, attached to the hump by a snake-like neck, looked up, and bright brown eyes set in wrinkled skin watched us keenly . . . with a slow deliberate movement he raised himself up in the mud and began to crawl away. . . . We heard a strange hissing sound. It was some moments before I realised that it was made by the tortoise as he slowly drew in his head. Like an escaping from a balloon, the noise continued, *hssss*, lasting for thirty seconds. . . . The high-domed shell wobbled away, supported by feet like cut-off stovepipes. He ploughed through the bush, dragging a long streamer of jungle vine that was hooked under his shell, crushing the grass, and leaving a wake of trampled verdure. His breathing could be heard clearly fifty feet away. . . . A vermilion flycatcher sat on a vine above him, watching him pass, and then darted off in a brilliant flash to catch an insect. It returned to perch on the slowly wobbling carapace of the moving tortoise.

Since the climate of the Galápagos is very different from

that of Aldabra, the daily routine of the tortoises on these islands is much more variable. Whereas on Aldabra they are already retreating into the shade from the heat of the sun at 9 o'clock in the morning, those on Indefatigable are only just becoming active at 8 or 9 o'clock, as the sun warms up. If the sun is shining from a clear sky they may return to the shade of the trees at noon, to sleep in a hollow at the base of a tree or bush; but on the many misty days, and during and after rain, they wander, feeding, to and fro along their trails.

Darwin was told by the natives that the tortoises remained for three or four days at a spring, before returning to their own territories. They can indeed store the surplus of water, not absorbed directly into the system, in a large sac in the carapace; but in dry regions they can no doubt survive for long periods, or perhaps permanently, on moisture and sap obtained from cactus pads and from the fleshy leaves of drought-resistant shrubs and plants. According to a native of Indefatigable, Gilbert Moncayo, who hunted tortoises for eighteen years, and subsequently marked more than 1,000 for the Charles Darwin research station on the 30,000 acres reservation at the west end of the island, some at any rate of the large highland tortoises move down towards the coastal regions in the dry season when the pads of the prickly cactus (*Opuntia*) fall to the ground, and one known to him took eight days to travel less than ten miles. Many are solitary, but on one occasion 280 tortoises assembled during a period of three weeks, in an area of standing water and manzaria trees 50 yards square. This was possibly a mating assembly, for some of the large males are known to descend for this purpose to transitional zones a mile or two from the coast, though never actually entering the arid region; but do so, however, in February and March, which is not the dry season.

It must be said that our knowledge of the giant tortoise's life-history is both sketchy and confused.

The smaller younger tortoises, and probably the females also, live mainly at lower levels. Perhaps the large adult males, which have been seen fighting among themselves and also harrying smaller tortoises, establish territories in the highlands. According to Moncayo, the females' nesting grounds are in the semi-arid prickly-pear zone and in the transitional areas, and there they lay their eggs in April or May. Darwin said that they dropped their eggs indiscriminately in any hole or rock fissure, but this seems highly improbable, and more recent reports suggest that they scoop out hollows in the sand or lava soil in sunny places, moistening them with urine in order to firm the sides before laying from five to ten or possibly twenty eggs. Then, according to Moncayo again, having pushed some soil over the eggs with her head, the female urinates on another heap and plasters this over the hollow, smoothing it and stamping it down, or pushing a stone over it. Moncayo followed one female for half a day and observed that after completing one laying and nest, she moved on a couple of hundred yards to excavate another hole and lay, and subsequently repeated this operation three more times; the first and last nests contained five eggs.

The puzzling fact is that the cap plastered over the nest dries hard, and there is evidence that, when they hatch after two months, the young tortoises—only 2 to 3½ inches long and 2 ounces in weight with soft enamel-white shells—are often unable to break through this cap, and die in the nest. This suggests that they can only break out after rain has softened the cap. Yet the eggs are apparently laid after the rainy season, such as it is, has ended; moreover the rains often fail, and in some areas are always slight, or negligible for years at a time. The only comment one can make on this apparent contradiction is that Moncayo, in confirming that some young tortoises are unable to break out because the cap has set like cement, states—if his interpreter, Charles C. Carpenter, understood him correctly—that those in a nest he fenced with a corral did not hatch until thirteen months later; moreover they did so when the soil was dry and their first food was restricted to grasses and fallen cactus pads.

We have seen that on Aldabra large numbers of young tortoises fall victims to predators during their first five years. On the Galápagos, however, their only natural enemies appear to be the hawks which, according to Darwin, take great numbers of them; though Moncayo refutes this. Actually, the principal food of this fine black buzzard-like raptor seems to have been the native rice-rat; but now that the latter has been almost ousted by the larger ship's black rat on all the islands except Narborough and Barrington it probably preys on the small *Tropidurus* lizards and a variety of birds. Though still breeding on eight of the islands, often nesting in a lone tree high on a hillside above a crater lake, the hawks are themselves decreasing, partly perhaps because of this decline in the numbers of their natural prey, but also from the guns of poultry keepers on the inhabited islands. But, as on all the islands in the Indian Ocean except Aldabra, man has taken a fearful toll of the tortoises on the Galápagos—a toll that began with the first post-Inca discoverers of the islands, the buccaneers of the Spanish Main. During the nineteenth century some 200,000 were shipped by whalers, 13,000 of which were taken by a fraction of the American whaling fleet out of New England during the years 1831 to 1867. Since the bulk of these were adult females, which were not too heavy to be carried considerable distances and which lived nearer to the coast than the males, it is surprising that sufficient numbers survived to maintain a breeding stock. Only the terrible nature of the terrain, fortunately exaggerated, can have saved the Galápagos tortoises from total extermination. They are indeed rarely found within a mile of the shore on any of the islands today, for there has been considerable human settlement, some intermittent, some permanent. Even in Darwin's time, the unique *Scalesia* forest on Charles, where tortoises are now extinct, was already being cleared for agricultural use; and although Indefatigable remained almost unexploited until the 1930s, settlers and their stock then infiltrated the virgin lands. By the 1960s only Narborough of the larger islands was still inviolate: not even a ship's rat had sneaked ashore. But Narborough is arid from its shores to the summit of its active volcano, and its tortoise population, probably never very large, has been decimated by eruptions. Beck saw a single live one during his ascent of the volcano some seventy years ago, and tracks were reported by the next expedition to the island in 1964; but an American zoologist, Paul Colinvaux, who climbed the volcano both before and after the severe 1968 eruption, saw no signs of tortoises.

The animals that man has introduced to the Galápagos—rats, dogs, pigs, goats and even donkeys—have also proved a scourge to the tortoises, particularly to their eggs and young. Rats, preying on the latter, may spell the doom of the colony of perhaps 150 on Duncan, before they can be exterminated, though this colony is now being reinforced by young tortoises reared at the research station. Donkeys break into the nest-hollows with their hooves, smash the eggs when rolling to rid themselves of parasites, and also eat down the prickly-pear. Beck encountered networks of donkey trails everywhere on James where the tortoise population has been reduced to single figures; and Shipton found that there were still hundreds of donkeys—in typically good condition and unusually large, as they are in the desert hinterland of Inagua—in the open woodlands around the 1,000 feet contour on Albemarle, while their trails even penetrated the dense thorn jungle on the crater rim of Volcan Alcedo. Both Darwin and Beck reported that immense numbers of young tortoises were eaten by feral dogs as soon as they broke out of their holes. Wild pigs dig up the eggs as soon as they are laid, and can also overturn well-grown young ones and eat their exposed fleshy parts. Pigs have been one of the main scourges on James and Indefatigable; but more than 1,500 have been shot in recent years and the numbers of young tortoises are increasing on Indefatigable, where the population is now estimated at about 3,000.

Since 1942 the tortoises have also had to contend with a succession of extremely dry rainy seasons, though periodic fluctuations in the rainfall have no doubt always been a feature of the Galápagos climate. But the effects of this latest long-term drought on the tortoises have been accentuated by the presence of goats, whose progenitiveness is so incredible that a he-goat and two nannies introduced to Abingdon in 1959 had founded a stock of between 3,000 and 5,000 by 1968. They have denuded the vegetation, especially on the more arid islands such as Hood, where only a dozen tortoises survive. Their habitat on the Galápagos may be as harsh as that of Aldabra. Nevertheless, even on such waterless wastes as Gardner, an islet off Charles, large numbers are able to maintain themselves in excellent condition by feeding, when all other vegetation has been eaten, on the bark of the tree cacti and on that dry lichen, *Usnera plicata*, whose nutritive value has been compared to that of steel shavings! Oddly enough, they do not, it is said, eat a common shrub, *Castela galapageia,* which the tortoises feed on extensively in the dry zone. Moisture they obtain by licking twigs and stones wetted by the fine mists, or from the sea itself.

8: Iguanas and Birds on the Galápagos

During his hundred hours on the Galápagos, between March and May, Beebe saw only one giant tortoise. For him, therefore, the interiors of the islands were dominated by land iguanas, while the jolly little *Tropidurus* lizards ran everywhere under foot; but the shores belonged to the big black marine iguanas which, more than any other creature he had seen in his uniquely wide experience, the hoatzin excepted, brought the far distant past vividly into the present. "His head was clad in rugged scales, black and charred, looking like the clinker piles of the island; along his back extended a line of long spikes, as if to skin of lava he had added a semblance of cactus."

The presence of these large marine iguanas on the Galápagos does not pose the same problems as that of the giant tortoises, for they are strong newt-like swimmers, progressing with regular rudder-like strokes of their long and laterally compressed tails—not, curiously enough, employing their webbed hind feet—and can also float motionless for several seconds in mid-water, without either rising or sinking. Moreover, like the tortoises, they can survive long periods of starvation; and some which Beebe took on board the *Arcturus* were strong and active after refusing food for a hundred days. A sea crossing of 600 miles from the South American mainland would therefore be well within their compass. On the other hand, their very ability to swim well does raise a problem. It was Beebe's belief that, though these iguanas were to be found on no fewer than twenty-three of the islands, there was only one valid species, in contrast to the multi-specific lizards and tortoises which could not, or did not, voluntarily take to the water. However, some zoologists now consider that every island race of iguana does exhibit some degree of sub-speciation. Those, for example, inhabiting the 6½ square miles of Tower's black lava-fields are coal-black pygmies, not exceeding 18 inches in length, whereas on other islands they grow to maximum lengths of 5½ feet and weights of 20 pounds. So too, on Hood, as burnt-up and arid as Tower but, as we have seen, with weathered red and also gray-blue and lavender-coloured slabs and boulders, the iguanas are startlingly different from the norm, with brilliant malachite-green dorsal combs, while their flanks are a vivid blood-red speckled with black.

This sub-speciation in a reptile obviously capable of commuting from one island to another, and which must, voluntarily or involuntarily, have been carried by currents from one to another on a number of occasions, is most puzzling. It is usually suggested that inter-island communication is prevented by predators such as sharks, and though no one has ever reported seeing iguanas with tails mutilated or missing as a result of escaping from attacks by predators, it is true that their remains have been found in sharks' bellies. It is also true that no one has ever encountered an iguana more than a few hundred yards out to sea from an island's shores: though how many people have been keeping a look-out for iguanas at sea off the Galápagos? Again, as Darwin first noted, and as other naturalists have subsequently confirmed, if an iguana is thrown into the surf, it immediately climbs out again on to the rocks. Those that Beebe experimented with in heavy surf dived instantly and, walking or crawling effortlessly along the sea bottom, clinging fast to the reef at every back-wash, gradually worked their way into deep submarine crevices that led to between-tide caverns; while one which he released from a boat floated for a minute and then, after diving three times to the bottom, finally swam, with head and much of its back above water, to the furthest point of lava 150 yards distant. That sharks may be one cause of the iguanas not apparently venturing into offshore waters is perhaps confirmed by Beebe's experience that it was just as difficult to drive them into the quiet waters off deep sandy beaches or beneath the ledges of low cliffs, because these were the waters frequented by the larger sharks which were not to be found at the edge of the breakers. However, although iguanas when ashore never attempt to escape capture by plunging into the sea, but go to ground in crevices or half way down the face of a cliff, they are often to be seen swimming from one lava promontory to another across the calm shallow waters of broad bays which are frequented by very small sharks.

But fear of predators does not appear to provide an adequate explanation for the iguanas' avoidance of offshore waters; nor does the suggestion that they cannot be driven into the surf because they are unable to withstand the smashing impact of the breakers. On the contrary, the zone of active surf is their marine habitat, for only in this zone can they obtain their almost exclusive food, the sargassum weed, despite the fact that they may be submerged beneath the incoming waves, and must cling fast to the rocks with their strong hooked claws.

> The waves break and foam around them [wrote Bryan Nelson] scouring the boulders and their industrious black gnomes, but the gnomes are still there when the water sucks back. Like the red crabs in whose company they so often feed, they can resist enormous suction pressures. Unlike the crabs they are none the worse if a heavy wave does dislodge them, for they wear their skeletons under their tough, scaly hides, nicely packed round with resilient tissues—difficult to break. One often sees a swimming iguana absorb a thunderous hammering on the stormy south-west coast of Hood, hurled violently against the rocks in the foaming water.

While it may be true that large sharks do not venture into the surf area, the iguanas exhibit none of that nervousness when entering their feeding waters which one would expect if they were regularly subjected to predation. Beebe never saw one actually dive in order to feed, for abundant supplies of seaweed were usually exposed on the rocks or on the offshore reefs. There are some 300 varieties of seaweed in the Galápagos, and no matter how large a particular colony of iguanas, there is a perpetual growth of weed in their feeding areas. But although the rocks are predominantly the grazing grounds of the smaller, younger iguanas, the adults feed mainly under water; and in areas such as the coast of Narborough, where the inshore rocks are not often exposed, or

where there are no reefs in 10 feet or so of water, the latter may venture 100 yards or more out to sea, in order to dive down to reefs 35 feet under water. They can probably feed at considerably greater depths, for they can remain submerged for more than half an hour at a time, ballasted perhaps by the stones they regularly swallow, in these waters pleasantly warmed to a temperature of 70 to 75 degrees F.

The Austrian zoologist, Irenäus Eibl-Eibesfeldt, has described how after spending the night in burrows, or deep down in crevices in the lava, the iguanas come abroad at 8 or 9 o'clock if the sun is shining, and then wait for the tide to ebb back from the rocks before making their way leisurely down to the edge of the surf, there to feed on the bared seaweed or to slip into the water and swim out to other weed-covered rocks. Crouching low on the slimy rocks, they apply their blunt, "sawn-off" muzzles to the weed's minute growth and crop it to right or left, or lay their heads sideways, like a dog with a bone, the better to graze with their sharp tricusped teeth, for the glutinous algae is tough and only the tips of the weed are eaten. They feed mainly during the hours of low tide, when there is less water over the reefs, and especially in the morning, before basking in the sun on rocks out of reach of the waves; but even with low night tides do not apparently feed in the evening or by moonlight. A diet of seaweed necessarily results in an excessive intake of salt; but the iguanas are able like some sea-birds to excrete the surplus through two salt glands opening into the nasal cavity, and periodically squirt this from their nostrils in fine jets of water that spray out in the wind like puffs of steam. This habit, together with their peculiar snorting, renders them even more dragon-like.

Although two species of lizards inhabit the tidal zone on Malpelo Island off Colombia and a gecko preys on tidal crabs, as does an iguanid on Cerralvo Island in the Gulf of California, none of these are as marine in their habits as the Galápagos iguanas, whose feeding habits are unique among reptiles, though they are not quite exclusively vegetarian, for the younger ones chase grasshoppers along the lava beaches and the adults have been seen to eat the placentas of sea lions. By and large, as Beebe said, the high tide marks the equator of their few yards of terrestrial and aquatic exploration. In this narrow zone they pass their entire lives on most of the islands, though on Tower Island (which, significantly, is not inhabited by their relatives, the land iguanas) they may climb up into the trees. The females are reputed to lay their two, or rarely three, eggs on the beaches at the end of the rainy season; though on Tower Island, where the whole of one sandy area was undermined by their burrows, Nelson watched them excavating in March, digging in the sand with their forefeet and kicking back the debris with their hind feet, though this was the middle of the rainy season, such as it was, on this dry and arid island. Covering the eggs with sand or soil, the females leave them to hatch; though on Hood, where suitable nesting terrain is at a premium, they are reported to guard their nests in order to prevent them being dug up by other excavating females. One is tempted to link this behaviour (if the report is correct) with the fact that on Hood the females assume at this time a conspicuous coloration, strikingly similar to the uniquely bright hues of the males; for Eibl-Eibesfeldt has described how, when competing for laying sites, the females are aggressive, and warning-off threat displays similar to those of the males may develop into head-pushing fights and then into actual vicious biting, resulting in broken toes.

For the greater part of the year the iguanas are notably gregarious, basking on the rocks in assemblies of hundreds, packed together and lying on top of one another. It is only at the approach of the breeding season that adult males take up territories of a few square yards of rock apiece and defend these (and perhaps their small harems of females) against intruding rivals. Eibl-Eibesfeldt has described how, if another male approaches a territory, the owner warns it off. Raising itself on stiff legs it struts to and fro, sideways on to the other, when its menacing jaws, opened wide, reveal a red mouth contrasting with the black face and nodding head and the hackles of the dorsal comb. If this threat display does not cause the intruder, displaying in a similar manner, to retreat, a ritualized form of combat develops, in which the two opponents rush towards each other, open-mouthed, but just before impact lower their heads and clash, when the horn-like scales with which their heads are armoured interlock. Each then tries to push the other out of the territory, and this trial of strength, interspersed with periods of threat display, may continue for an hour or so until one or the other is eventually forced out. The loser then assumes a submissive posture, in which it lies on its belly, while the victor waits for it to retreat. This it does by crawling backwards, belly to ground, but still nodding its head, until it has retired to the proper distance, when it raises itself off the ground and struts away in full display panoply once more.

It is not necessary to enlarge upon the advantages of this ritualized combat, common to reptiles, mammals, birds and fish alike; but if it is to be conducted along recognized traditional lines it must be preceded by the introductory mutual display. Until they have learned this ritual, young iguanas bite each other, while if an adult male is introduced into another male's territory, without having first been allowed to conduct the traditional preliminaries, the two fight in earnest.

Outside the breeding season the iguanas are the most phlegmatic of animals, displaying that immobility and total indifference to external phenomena characteristic of reptiles. They crawl haphazardly over the backs of sea lions hauled out on the rocks and, in turn, with equal unconcern, allow the large scarlet crabs to crawl over them. The latter, our old friends the Sally Lightfoots, are an ever-present feature of the Galápagos beaches, with as many as 500 painting the black lava and pink strata of a small cove with violent splashes of colour. Beebe has put on record his astonishment on witnessing one of these crabs slowly approach a 3-foot-long iguana, which was resting partly on the sand and partly on a water-smoothed stone, and instead of turning aside on reaching the iguana's head, climb directly over it, causing the lizard to close its eyes to protect them from the crab's sharp legs. While crawling along the whole length of the iguana, the crab stopped three times in order to nip off a tick from the latter's skin; but although this involved considerable tugging on the part of the crab, with the result that the iguana's skin was pulled high above its back, the latter did not object. On subsequently examining the iguana, Beebe found that the crab had not removed sixteen ticks, which proved to be closely related to those parasiting the land iguanas.

Whether or not iguanas are preyed on by sharks and other marine predators, they do not evince the slightest sign of fearing any land predators or of ever having had to protect themselves against attack. When harried by man they neither bite nor scratch, nor butt with their heads, nor even erect their dorsal spines threateningly.

> Here [observed Beebe] are great lizards with mail of scales, which become like solid masonry around the head, with a formidable saw-like ridge of horny teeth down the back and tail, with many small but efficient teeth and powerful jaws, with twenty long curved talons, backed by incredibly strong muscles and yet with no desire or power of defence or lashing with the tail—much less dangerous to pick up than the big scarlet crabs.

In the absence of definite evidence to the contrary, then, one must reject the usual assertion that they have one natural predator—the maligned Galápagos hawk. All the evidence is that the iguanas are in fact as unconcerned by the presence of a hawk as by a crab. Neither Beebe nor Nelson witnessed any attack by hawks, and considered that iguanas are too large and too well armoured with mail and powerful talons to form their prey, though the land iguana, which is also armoured and reaches a length of 3 feet, is reputed to repel attacks from hawks by threatening them with open jaws.

In the presence of man both species of iguana are almost fearless, and in remoter districts can be approached within a few feet. Beebe indeed claimed that the marine iguanas could always be touched and even stroked; but today, in the vicinity of settlements, they have become shy as the result of persecution, though considerably less shy than their South American relatives, which dash into cover at sight of man. This fearlessness is common to all wildlife on the Galápagos though relative perhaps in the case of the giant tortoises, for these are apparently deaf, evincing no sign of hearing the human voice at a range of a couple of feet. Only when one steps close in front of a tortoise does it utter a loud hiss and perhaps withdraw its head beneath its carapace. Couffer was accompanied for miles by three of the Galápagos hawks, which continually perched at arm's length from him; and he describes how when he and his companions were camped on a beach at dusk in an orange canopy of firelight, a soft flitting shadow cast by the starlight made him turn, and he found himself looking into the wide unblinking eyes of an owl, standing on the beach not 8 feet from him:

In a moment, another shadow glided very swiftly across the

> sand. On wings as soft and silent as the filtered starlight, a second brown owl dropped to the sand beside the first. Others followed, perching on the rocks and cactus branches, or on the beach. They stood on the edge of the firelight, watching us. . . . Eventually there were eleven around our campfire . . . watching us in the same way that the night herons had done at other camps. . . . The owls stayed on the beach that night, watching us until we had finished our meal and climbed into the skiff to row to the boat. Then . . . they took wing and followed us, hovering over our heads as they circled the slowly moving skiff. . . . When we climbed aboard the ketch, all eleven of the owls found perches in the rigging.

It is generally claimed that this fearlessness of the Galápagos fauna and avifauna can be attributed to the fact that they have never been preyed upon by indigenous mammalian predators; but since there have now been predatory dogs, cats, rats, and pigs on the islands for some centuries, this argument is obviously not valid. Moreover the small passerines, which are the tamest of all so far as man is concerned, have always been extensively hunted by short-eared owls which are resident on almost all the main islands, and to a lesser degree by the hawks and barn owls; but the latter's principal prey was, as in the case of the hawks, the rice-rat, and these owls are now rare. Yet the passerines by no means always display fear of these traditional predators. Fearlessness is a notable feature of the wildlife on islands in many parts of the world. Gilbert Klingel found the birds of Great Inagua, sparsely inhabited by humans, to be completely tame; while those on the Falkland Islands were formerly also tame in the presence of man, despite such indigenous predators as hawks, owls and foxes, but became less tame after the islands had been permanently settled by man. This is a fascinating subject, but a comprehensive enquiry into its various aspects would require the full treatment of a thesis and cannot be pursued here.

Since it is the innocent who perish, the universal tameness of the Galápagos wildlife has rendered them particularly vulnerable not only to man but also to such introduced predators as dogs and pigs. Thus, even in Beebe's day the marine iguanas were slowly but certainly decreasing in numbers. Today, though still clustering in hundreds on the rocks of hostile Narborough, they have become rare near settlements on such islands as Chatham and Charles. Free from disease and natural predators, and with access to perpetually abundant sources of food, these iguanas ought to have been allowed to decline peacefully into old age until, as Nelson has imagined it, "One day, in the full heat of the Galápagos sun, with the black lava at a temperature of about 140°F, the ancient lizard, as black and craggy as the rock, clambers lingeringly to his favourite slab. Nodding even more slowly, he directs the last derisive vapour puffs from his nostrils. Then with 'his smug lizard's smile unchanging' he returns to the womb of the island that spawned him."

Beebe has said that the marine iguanas, though terrestrial in origin, were obliged to adapt themselves to a littoral and marine habitat on the Galápagos because of the barren nature of the lava deserts that form the periphery of so many of the islands. Possibly: but there is also the alternative that the pioneer immigrants to the islands had in fact come from littoral habitats. However, some found terrestrial life to their liking in the arid coastal regions, though comparatively few penetrated the damp highlands, and eventually evolved a distinct land species, *Conolophus* (the marine iguana is *Amblyrhynchus*). In contrast to all the races of marine iguanas except Hood's, *Conolophus* glows with colour against the sparse dark-green vegetation. Its great head, with small shining orange-red eyes, is rugged with rough, bright golden scales; and these small cusped armour-plates, resembling clusters of crystal, form a green mosaic on the lips and are chrome-yellow on the underside of the chin. Body and tail are partitioned into large irregular sections variously coloured terra-cotta or brick-red, black and an intense canary-yellow; while the dorsal comb consists of stout horny spikes, diminishing in size towards the tail, and coloured according to the section they spring from.

Amblyrhynchus can swim strongly, and must have commuted from island to island on occasions: yet has apparently evolved a number of different races in isolation. *Conolophus*, on the other hand, is terrestrial and, though it can certainly float, is reputed, improbably, not to be able to swim. Nevertheless, colonies have been established on eight of the islands, derived perhaps from original stocks of marine iguanas; but though they have evolved in, one would have supposed, conditions of much more restricted isolation than the latter, only one sub-species exists—a pale-coloured race on Narborough. There do not seem to be any circumstances that can explain this paradox; but there is no doubting that these terrestrial iguanas were extremely successful in their arid habitat. On James, for example, they were so numerous that Darwin was unable to find space to pitch his tent because the ground was undermined with their burrows; while on the savanna of Seymour, an islet to the north of Indefatigable, Beck observed that every cactus, every isolated small bush of *Cordia* or *Acacia* or fragrant *Bursera* provided shelter for a *Conolophus* and that none was less than 2 feet long. But while the majority remained in the coastal desert lands, it is significant that on Narborough, where arid conditions prevail from sea-level to mountain top, they dispersed to the very summit of the 5,000-feet-high volcano. Colinvaux, who climbed the volcano both before and after the 1968 eruption, has described *Conolophus*' habitat on Narborough. He found that the lava wastes of large cinders, waterless and plantless, with not even a cactus, reached down to within a few yards of the coastal strip with its abundant life of marine iguanas and sea lions, penguins, pelicans and flightless cormorants. Above the cinders area was a shifting jagged scree of tumbled blocks, succeeded by strips of battleship-gray plate-lava in the form of solidified flows, swirls and flat-topped tubes; then steep screes of 3-foot blocks again, and the first signs of life in the shape of cacti and a light brown and yellow *Conolophus* climbing slowly away over the scorching-hot waste of lava, on which were the bleached bones of two others. When 1,000 feet or so below the summit he reached the zone of intermittent low cloud, and here, although the ground was still parched, there were dried grasses and ferns growing from cracks in the lava and tall cacti with lemon-shaped fruits. Another area of scree up to the craggy ridge brought him to Narborough's only relatively lush greenery in the form of a forest of low, shrubby and dusty *Scalesia* trees with straight trunks a few inches thick and curious tangles of branches, beginning at head height and bearing rosettes of leaves capped with large dandelion-like flowers. Among the trees were clumps of brown hay-like mats of grass higher than a man, and clouds of choking dust. This was the inviolate sanctuary of the *Conolophus* (and some cane-rats) basking in the sun and, when almost within touching distance, dashing off into the bushes or down the burrows which undermined the sandy flats.

After travelling through the forest for almost three-quarters of an hour, Colinvaux and his companion reached the rim of the crater, elliptical in shape and 12 miles in circumference, and descended the 1,800 feet of its mainly sheer walls into the interior. Here too the sandy floor in the 3-mile-long pit of the crater was a warren of iguana burrows. Cane-rats were also present, and a hawk was perched on a pinnacle of lava. A sweet-water lake, a mile long, at one end of the crater was the remarkable home of some 2,000 pintail duck. These, possibly, had been responsible for stocking the lake with small fish, for, improbable though it must seem, there are documented records of birds fortuitously transporting fish in their plumage. (In like manner, the two species of earthworms in the highlands of Indefatigable could have originated from eggs transported on the feet of birds.) There were also two smaller lakes in secondary cones within the crater, whose waters were as warm as those of a hot bath and highly sulphurous; and from the slopes of one of these cones Colinvaux collected thirty species of plants.

A month after the major eruption Colinvaux again climbed to the crater, but by a different route. On this occasion he followed a long flow-strip of gray plate-lava which stretched unbroken from the coastal bluffs to the base of the ultimate steep ascent, and this route was, significantly, vegetated. Although it was a desert vegetation, there were *Scalesia* bushes and *Bursera* trees and, from the outset, animal life, including *Tropidurus* lizards, a pelican nesting in a bush, a hawk, and doves in the leafless gray-barked *Bursera* trees, while warrens of iguanas undermined sandy patches filling hollows in the lava flow. Although there was 3,000 feet of steep scree—an immense hill of slag—above the flow, this vegetated corridor could have been one route by which the iguanas had dispersed to the summit of the volcano.

The astonishing feature was that, despite the violence of the eruption, there had been no great changes to the exterior of the volcano, and there was little evidence of its blast except for rifts a yard wide and a heavy deposit of gray dust on flowers and bushes and on the tendrils of the clinging legumes. Iguanas still inhabited the *Scalesia* forest, doves and finches visited the camp, and a pair of owls hovered over it after dark. While the plateau round the rim of the crater had been little more than dusted, the full effects of the explosion had been restricted to the interior, one end of the floor of which had been blown out and superseded by a chasm perhaps 500 feet deep and a mile across. The small cone, from which Colinvaux had collected his plants, had slipped like a block of cement to the bottom of the chasm, and the tiny lake in its crater had merged with the large lake, which had shifted more than a mile from its original position. One must presume that, with the possible exception of some of the duck, the original fauna in the crater must have been destroyed; yet a few iguanas, lizards and geckos had already recolonized it less than three weeks after the eruption. One of the world's few remaining virgin natural laboratories could begin working again.

The success of these iguanas in exploiting a habitat niche in the arid coastal zone of other islands, rather than in the luxuriantly vegetated highlands, must be due to the fact that, like the tortoises, they adapted themselves to obtaining adequate supplies of food and water from the cacti. It is true that they consume enormous quantities of plant leaves and flowers, especially the large yellow tubular flowers of *Cordia* and the yellow petals of the puncture-vine *Tribulus*, together with the shoots and bark of such shrubs and trees as guavas and stunted acacias, but these are only available for a few months in the year, and their basic food is the cactus fruit.

They are reported to scrape the larger spines off the cactus pads with their claws before eating the latter; but Beebe, having extracted five whole pads from the stomach of one *Conolophus*, and also having observed that their droppings were composed of masses of the needle-length, steel-hard spines, was puzzled by their ability to swallow these without harmful effect. *Conolophus* is indeed an important ecological asset, for from its droppings, in arid regions and highlands alike, sprout vigorous colonies of seedlings.

Alas these splendid dragons, and especially their young, have proved even more vulnerable to the depredations of introduced domestic predators than their marine relatives. On James and Seymour they are now extinct. Rats and pigs destroyed their eggs, dogs killed the adults, and finally 3,000 American GIs, based on Seymour during World War II, exterminated the entire fauna on that island. They are now rare in most regions of Indefatigable, except the north-west, and in the neighbourhood of settlements on Chatham and Charles; while although there are still fair numbers on Barrington, very few are young, possibly because the goats, before their extermination in 1971, by eating down the shrubs and bushes deprived them of escape cover. Only on Narborough does *Conolophus* continue to thrive.

The sub-speciation of the marine iguanas is paralleled by that of certain birds of the islands. Large colonies of sea-birds breed on the Galápagos. There are some 280,000 red-footed boobies on Tower alone. Thousands of storm petrels nest in burrows under the cinder plains; precipitous slopes are dotted with white boobies and riddled with the holes of the red-billed tropic-birds; blue-footed boobies nest in Beck's "broiling and insufferable ancient craters." There are the unique flightless cormorants and the most northerly colonies of penguins.

But it is the small passerines that offer the most extraordinary evidence, by their fascinating adaptations to their unique environment, not only of the Galápagos' insularity but of island sub-speciation. Yet, since all have retained their powers of flight, there would, one would have thought, have been frequent traffic among the islands, if not with the mainland

Conolophus, *or land iguana*

of South America; but in fact there may be as many as 10 sub-species on a single island of the central group; while 77 of the 89 species and sub-species are found nowhere else in the world.

This sub-speciation is not, however, peculiar to the Galápagos. On the much more isolated islands of the Hawaiian Archipelago for example—equidistant 2,000 miles from California to the east, Alaska to the north, and Japan to the west—twenty-two kinds of honeycreepers have evolved from a single passerine ancestor, possibly of American origin. When a new species of bird establishes itself on an island it may find a variety of suitable unfilled habitat niches available for exploitation; but as its descendants multiply, competition may oblige them to disperse in search of food and nesting sites into untypical habitats. In these, only those individuals able to adapt to the unfamiliar conditions, because of some physical peculiarity, survive, and transmit these peculiarities to their progeny. Eventually, from an ancestor which perhaps resembled a sparrow, are evolved sub-species with brush-like tips to tubular tongues for extraction of nectar and also pollen (which may possibly include a vitamin not present in nectar); sub-species with curved beaks with which to tweezer out spiders and small insects from bark and moss and the stems of tree-ferns; and others with parrot beaks that can crush large seeds and fruits.

Many years ago the American naturalist, Frank M. Chapman, summed up the possible origins of sub-speciation most succinctly when considering the birds of the Bahamas, where different forms of the same species inhabit islands almost within sight of each other, and whose climate, soil and flora are essentially similar. He suggested that:

Perhaps we can assume that through the continued association of a comparatively small number of individuals, certain characters, due originally purely to individual variations, have become perpetuated and specific. Among a smaller number of birds the extent of variation would not be so wide; but this would be counterbalanced by the fact that any dominant character would be far more likely to be preserved through the forced interbreeding of closely related individuals. This would also hasten the consummation of permanent forms; the rate of divergence among island-inhabiting species being, therefore, more rapid than among those of the mainland.

The most fascinating feature of this sub-speciation is the variety of ways in which birds have been forced to adapt to untypical environments, and to experiment with and perhaps take advantage of unfamiliar habitat niches not already occupied. No doubt the general shortage of water in the more arid regions of the Galápagos has presented a problem to those birds unable to obtain moisture from cactus pads and the fleshy leaves of shrubs. How serious this problem may be is illustrated by Bryan Nelson when describing the habits of mocking birds, which are rather smaller than blackbirds and weak fliers; but they are spare and agile, with long beaks and legs on which they run very fast with wings outstretched like ostriches. On the very dry islands of Hood and Tower the mocking birds would flock enthusiastically to the stinking contents of ancient albatross eggs, probably more than a year old, which Nelson broke for them, just as they would collect around an adult albatross feeding its chick and scrape up any oil that dropped from her beak. They were also passionately fond of fruit, swallowed lumps of beetroot, and drank vinegar as though it were nectar. However, as Nelson expressed it, pride of place must surely go to those which drank a warm and concentrated solution of Dettol; although shaking their heads violently after every sip, they still returned for more.

In more serious vein, Nelson observed a mocking bird dart beneath the tail of a blue-footed booby and peck at blood on its cloaca, actually lacerating the tissue and pulling off fragments. On Wenman—a small north-westerly outlier—ground-finches have been reported behaving in a similar manner, and have also been seen to peck at the wing-elbows of masked boobies and drink the blood that flowed; moreover, if one of these finches was shot, several others would congregate to sip at its blood.

There is nothing particularly noteworthy about vermilion flycatchers employing the backs of tortoises as aircraft-carriers from which to launch out in pursuit of insects; nor of finches alighting on the heads of tortoises in order to remove minute grass seeds from their nostrils and the corners of their mouths, or diving into the opening between neck and carapace in order to pick at the mites and ticks lodged in the folds of the tortoises' skin. There are many parallels in the world of birds to this behaviour. A more unusual food niche has, however, been discovered by the yellow warblers on Tower. Nelson, again, has described how these colourful little warblers with crimson crown feathers were often to be seen stalking the tiny fiddler crabs, a dense colony of which lived in a sandy creek. When the fiddlers emerged from their burrows, after an ebbing tide had deposited a new supply of detritus at their doors, the warblers would flutter here and there, snapping feverishly at the crabs. Although these withdrew into their burrows in a twinkling the warblers' persistence enabled them to catch several and then dismember them by dashing them against the ground, before swallowing the fragments.

But consider the diversity of beaks evolved by Galápagos finches. Of three ground-finches feeding mainly on seeds,

Woodpecker finch using cactus spine to pry grub from a hole in a log

one with a beak as massive as a hawfinch's crushes hard seeds, a second with a much smaller beak feeds on small seeds, and a third with a medium-sized beak on intermediate seeds. The large tree-finch's decurved beak enables it to feed on flowers in the manner of a small parrot, while the cactus finch has developed a strong pointed beak for coping with the fruit and pulp of the cactus pads. But the most remarkable adaptation of all has been achieved by the woodpecker-finch, which has developed a stout straight beak comparable to that of its namesake. With this it not only explores vertical trunks and branches in characteristic woodpecker manner, but also chisels into their boles in order to obtain wood-boring beetles and other insects. However, since its tongue is not long like a woodpecker's it cannot be inserted into a crack, or into a hole the bird has excavated, in order to pick out an insect;

but this remarkable finch is not defeated by this handicap. Selecting a needle-long rigid spine of prickly-pear if it is in an arid zone, or breaking off a twig, an inch or two long, from bush or tree if it is in a humid zone lacking cacti, and holding this tool lengthwise in its beak, the bird either uses it as a crowbar to pry off flakes of bark from the rotten trunk of a dead tree, testing borings for grubs, or pokes it into a crack and winkles out an insect, dropping the tool when its prey emerges. It may even carry the tool from one tree to another and probe a number of cracks with it. This in itself is an extraordinary development, paralleled only by the bower-bird, *Ptilonorhynchus viloaceus*, which employs the juices of fruits to colour the stems with which it strews its mating bower. But the woodpecker finch is also capable of rejecting a twig that is too short or too pliable, and of caching a store of spines for subsequent use. Possibly the mangrove finch, which has a similar type of beak and feeds almost exclusively on insects obtained by probing under the bark of trees, also makes use of tools; but little is known about the habits of this finch.

There is a danger, of course, that such ingenious but highly specialized adaptations could prove a fatal handicap in the event of any sudden change in habitat conditions. This has not yet proved to be the case on the Galápagos; but on the Hawaiian Islands such features of encroaching civilization as housing development, the draining of marshes and the felling of forests to make way for plantations of sugar-cane and other crops, have resulted in the extinction of 8 of the 22 honeycreepers, while a further 4 or 5 survive only in remote regions.

9: Dragons Large and Small

"If we had been alive in the Age of Reptiles the great sight would have been the sweep of the deployment: the adaptive radiation of unprecedented types of life into unprecedented ways of living," wrote Archie Carr in *The Reptiles.*

> Today, looking at what remains, we see the lesson of extinction and survival. The whole spirit of the grand, doomed days seems held in the small, cold body of the *tuatara* on the few chill islets of its dwindled range. Living on as unaccountably as the giants it once lived with died, it creeps out of its burrow in the evening, plods about in the mist to gather crickets to keep the cold, small frame of its life alive; then, in the early morning, goes back again into the earth to which it has clung so stubbornly for some 200 million years.

A stocky large-headed lizard-like reptile with a short toothed crest of white spines or folds of skin along neck and back—hence *tuatara*, the Maori term for "peaks on the back"

Tuatara *emerges from its burrow*

—the male is only about 2 feet long and a couple of pounds in weight, the female slightly smaller. Neither turtle nor crocodile, lizard or snake, the tuatara is assigned to its own special group of reptiles, the *Rhynchocephalia.* Its small saurian ancestors, which are not known to have exceeded 5 or 6 feet in length, preceded the dinosaurs by perhaps 30 million years, and it, the last survivor of its group, has outlived them by 60 million years. A few hundred years ago the

Maoris, though fearing the tuataras, were eating them in ceremonial feasts on the mainland of New Zealand, but today these reptiles are restricted to twenty or more small islands off the North and South Islands.

The long-term survival of this small population of reptiles over such an immense period of time is inexplicable, though perhaps the tuatara's very low metabolic rate has contributed to this, for its body temperature of 52 degrees F is lower than that recorded for any other reptile; moreover it is reported to breathe only once every seven seconds and to be able to retain a single lungs-ful of air for as long as an hour. It seems likely that individuals may live for more than 100 years, and possibly to 150; one has certainly survived for 77 years in captivity, and there is a Maori remembrance of another attaining 300 years. But against the advantage of longevity must be set the disadvantage that the tuatara probably does not mature until more than 20 years old, while its 8 to 14 eggs, laid in a shallow depression in friable soil, do not hatch for 12 or 14 months. However, in the short term, its survival has been made possible by the large numbers of sea-birds roosting and breeding on the islands it inhabits. With these the tuataras live in intimate association, for though capable of digging out their own burrows they regularly make use of those excavated by nesting petrels and shearwaters, sometimes moving in while these are still incubating or rearing chicks, and possibly killing the occasional chick or adult—though whether the tuatara always occupies the right side of the burrow and the bird the left, as reported, is another matter. The survival of the tuataras is indeed a beautiful example of the value of ecological relationships. The multitudes of roosting and nesting petrels, shearwaters, terns, cormorants, and penguins break down the under-brush of the salt-forest and manure the ground heavily with their odorous guano, providing an ideal environment for beetles, snails and especially the large ground wetas or wingless crickets on which the tuataras feed, though they also prey on geckos and skinks. For these they hunt nocturnally, even when the temperature is as low as 45 degrees F; but during the winter hibernate in their burrows, though basking at the entrances on sunny days and for hours at a time in spring and summer. In the remote sanctuaries of these small islands the tuataras have become almost immortal.

In eastern Indonesia on the three small islands of Komodo, Padar and Rintjan and the two islets of Owadi Sami and Gili Moto, and along the west and north-west coasts of the large island of Flores, and nowhere else in the world, another small population of ancient but, by contrast, gigantic reptiles has survived—the Komodo dragons or oras, whose gigantism is, like that of the tortoises of the Galápagos and the Seychelles, associated with a restricted living space. The males of these varans or monitor lizards grow to lengths of at least 10 feet and weigh 200 or 250 pounds, and the females to 7 or 8 feet. David Attenborough, the British television wildlife photographer, has described a very large specimen:

> From the tip of his narrow head to the end of his long keeled tail he measured a full 12 feet. . . . I could distinguish every beady scale in his hoary black skin, which . . . hung in long horizontal folds on his flanks and was puckered and wrinkled round his powerful neck. He was standing high on his four bowed legs, his heavy body lifted clear of the ground, his head erect and menacing. The line of his savage mouth curved upwards in a fixed sardonic grin and from between his half-closed jaws an enormous yellow-pink forked tongue slid in and out.

The oras' islands are, like Krakatau, a part of that colossal volcanic chain extending south and east from Sumatra, through Java, Bali and Flores, and then northwards to the Philippines; and like the Galápagos they are of comparatively recent origin. The few surviving oras are probably descendants of even larger lizards, fossilized remains of which in Australia date back perhaps 60 millions years. But if they are the descendants of giants, how do they come to be on these geologically recent uplifted islands, and nowhere else? Driven out of their original habitat by some environmental crisis, they perhaps emigrated north-westwards from Australia by way of the Sahul Shelf which extends almost to Timor, and then step by step along the old arc of volcanic islands to Soemba and finally Komodo, bridging the gaps between the islands with the aid of the south-east monsoon currents. It is true that these Indonesian islands, like those of the Galápagos, are separated by exceptionally strong tide-rips racing through the straits at 13 knots when flood-tides are backed by south-easterly winds; but the immensely strong oras swim easily, if awkwardly, with heads held well above water, and can remain under for 4 or 5 minutes and probably longer. Indeed Walter Auffenberg, who undertook a year's field study of the oras on Komodo and Padar has reported that, when food is scarce on Komodo, large oras swim across the 500 yards of tide-race to the islet of Nusu Mbarapu in order to prey on the goats, with which most of these Lesser Sunda islands are periodically stocked, and remain there until these have been too thinned out to be hunted profitably.

It is little more than sixty years since the existence of the oras was first made known to the outside world by pearl-fishers, and subsequently by collectors from Java's Buitenzorg Zoological Museum who obtained five specimens in 1912. Today, rather more than 1,000—perhaps one-fifth of the total population—are estimated to inhabit Komodo, which is 22 miles long by 12 miles broad. Komodo is a dry island, with a rainy season extending only from late December to early March. By April most of the valley streams have ceased to flow, and the freshwater pools in the rugged, castellated mountains, less than 2,500 feet high, also dry up as the season progresses. Auffenberg has described how by July the leaflets of the tamarind trees are falling in rust-coloured showers; while by August the island's vegetation is gray and withered, with surface temperatures reaching 167 degrees F in the midday sun, and the small herds of deer and hogs, seeking water, churn up the powdery layer of volcanic ash into miniature dust-storms. In such conditions forest and thorn scrub are mainly restricted to the upper reaches of the hills where rainfall is more frequent, and thinly to the low parts of the valleys and ravines, whose dry stream-beds contain some sub-surface water. The wooded parts of the latter are also less vulnerable to the islanders' seasonal burning, which is gradually decimating the trees—notably tamarinds and the great gubbong or lontar palms whose lofty bare boles are crowned by Catherine-wheels of spiny fronds—with the result that the dominant vegetation clothing the rough and eroded surface of volcanic rubble is the alang-alang grass, growing not only at the mouths of the valleys, but on flat plateaux on the lower slopes of the hills, and sweeping waist-high over all the ridges dividing the deep valleys and ravines.

The latter, and the forested slopes of the central mountain ridge, are the oras' haunts; but our knowledge of their habits is both sparse and confused, or was until Auffenberg's study in 1969–70, to which I am heavily indebted. By one traveller we are told that they do not venture out of their "earths" in the rocks or under tangles of overhanging tree roots until

9 A.M., when it is beginning to get really hot: by another that they are slightly more active in the early morning and late afternoon than during the midday heat; but it seems reasonable to assume that they avoid excessive dehydration in the extreme noon heat by remaining in the shade of the denser groves of trees or in their earths, each of which has a broad platform cleared of vegetation at its entrance. An 8½-foot male, most familiar to Auffenberg among the fifty oras inhabiting the 4 square miles of his control area, would lie up under a hillside clump of shrubs just within the forest edge. Although the air temperature outside the thicket might be 122 degrees F in the afternoon, it was some 40 degrees cooler within the shade; and this male, like other oras, used a number of such lairs, which were invariably exposed to the breeze. At 8 A.M., when the sun had warmed the hillside, he would set out in search of food, working his way slowly down from the crags, stalking through the 4-foot-high alang-alang grass, with magnificent head swinging heavily from side to side and forked yellow tongue, 18 inches long, incessantly flickering in and out of his immense jaws, probing for scent of prey or carrion. At 5.30 P.M. he would retreat for the night to temporary lair or permanent earth.

It is possible that their present habitat could not support more than the present population of these gigantic varans, and Auffenberg indeed believes them to be as numerous today as they have ever been; but the thought does cross one's mind that their future would be more secure if they were normal-sized monitors, twenty-four species of which are widely distributed throughout southern Asia. Obtaining sufficient food to maintain their great bodies must be a problem, and Auffenberg refers to one large male which had almost succumbed to starvation at the time of his arrival on Komodo, but which survived by feeding on the expedition's baits and subsequently killing a hog. Prior to Auffenberg's researches it was assumed, with reason, that these ponderous reptiles must subsist largely on dead and crippled game, which they were reputed to scythe down with their long muscular tails, though the only unimpeachable evidence that live prey might be captured was provided by a solitary photograph of a Rintjan ora in the act of swallowing a live but paralysed macaque; and this incident suggested that an ora may induce in monkeys the same nervous paralysis that afflicts them in the presence of large snakes and, apparently, jaguars and tigers. There was also evidence that on Padar, which is a barren coral island holding very little game, the oras dig up the eggs of turtles, as they do those in the huge nest-mounds of the brush-turkeys or megapodes. But the bulk of their food was presumed to comprise carrion, though the inhabitants of the Lesser Sundas were well aware that tethered goats could not be left unattended on the savannas for any length of time before being killed by oras. For that matter even the presence of barking dogs and shouting humans did not inhibit oras from attacking goats tethered beneath their owners' stilted houses. Every photographer has reported the infallible attraction to oras of putrid bait, and Auffenberg proved that after 48 hours' decomposition a carcass would draw numbers of them from a large area, some from as far distant as 5 miles. In addition to their primary value as food, decomposing carcasses are important in the oras' biology as focal points for their social life and mating. Douglas Burden, who visited Komodo in 1926 as a collector for the American Museum of Natural History, has observed how:

For several minutes the observer in the boma [hide] might see

The Komodo dragon

> absolutely nothing, and then suddenly, from behind a tree, a big black head with two dark beady eyes would appear and remain entirely motionless, while the eagle eyes, sunken grimly beneath projecting supra-orbital ridges, would survey every inch of ground. . . . When the beast was assured that all was well, he would lower his head, flash a long yellow bifurcated tongue into the air, and then move rapidly towards the bait . . . the long sharp claws are used indiscriminately to scrape and tear with, while the thin, recurved teeth with serrated edges are employed to rip off great chunks of the foul meat. The beast manoeuvres this by see-sawing back and forth on broad legs, giving a wrench at the bait with every backward move. . . . When a piece of flesh has been detached, he lifts his head and gulps down the whole slab. . . . As the food goes down, the skin of the neck becomes distended in the most astonishing fashion, since the jaws, like those of snakes, are inarticulated, while the brain is protected from excessive pressure by a bony plate.

But since the only other large predators on Komodo are the recently arrived feral dogs, which compete with the smaller oras for carrion, and since both pig and deer are taboo to the few Mohammedan inhabitants, how could sufficient numbers of game be killed or disabled by other agents to provide carrion for more than 1,000 oras? There appeared to be no answer to this question until Auffenberg began his research. We now know that for the first year of its life an ora is purely a predatory lizard, slender bodied, thin-tailed, most of whose time is passed in the trees hunting such small animals as geckos. Not until it is about one year old and 3 feet long does it begin scavenging on the ground for carrion. At this age it is much more brightly marked than the uniformly black or sepia-coloured adult, with pale spots on its grayish-black back, dull orange flecks on shoulders and fore-limbs, orange throat and belly, and alternate bands of light-gray and black round the tail. But Auffenberg discovered that oras, far from being exclusively carrion eaters, were successful predators on almost every animal in their habitat. Very small ones subsisted mainly on insects; medium-sized ones on birds and rodents; large adults on, in particular, the Timor or rusa deer whose stags may reach 450 pounds in weight, and also on wild pig, goats, feral horses averaging 550 pounds, and even the massive water-buffalo weighing well over half a ton and a formidable prey for a tiger, let alone a lizard. Men have also been killed when sleeping on the ground during the daytime or working in the bush. Large oras also kill and eat smaller individuals of their own kind, especially when their numbers are at a peak, exceeding the optimum of perhaps twenty-five to the square mile of savanna grassland.

But how are the slow-moving adult oras able to capture and kill the typical prey of the swift tiger? Primarily by employing their exact knowledge of the local terrain and of the daily routine of its other inhabitants. Pigs, intent upon their rooting along the trails for fallen tamarind fruits, can be stalked to within a few feet; deer, which follow a number of regular trails, can be ambushed by an ora lying in wait for two or three hours in the bush or beneath a log. If an unsuspecting deer passes within 3 or 4 feet of a waiting ora the latter can lunge forward and clamp its massive jaws and recurved teeth on leg or throat or head and, having pulled the deer down, tear out its intestines. Horses can be ambushed on the trails too, and are also particularly vulnerable when asleep on the ground or when foaling, for an exhausted mare can do little to prevent an ora from seizing her 45 or 55 pound foal as she drops it. Water buffalo are too powerful and dangerous to be tackled directly, but can (with the tiger's technique) be hamstrung and, when brought to the ground, eviscerated; though most buffalo "kills" take the form of carrion that have died from other causes. Even if a victim breaks loose from an ora's grip it may still become carrion eventually, for wounds inflicted by oras tend to become septic and result in massive infection.

The chances of a deer or pig passing along a particular trail an ora has ambushed on a particular morning are, of course, fairly slight, and Auffenberg suspected that an adult ora might perhaps make a successful kill or locate a large dead beast only once a month. The latter it would probably have to share with other oras, large and small. However, if on its own, the kill or carrion could be disposed of in a matter of minutes, for a large ora can swallow the whole head of a wild boar or the entire hindquarters of a deer at a single gulp, and consume the 90-pound carcass of a hog in 17 minutes. If the morning's ambush produces no results the ora can try its luck later in the day hunting for rats and jungle-fowl or searching methodically through the thick covert in which deer and pigs lie up. Peering intently into each bush for 10 or 20 minutes before passing on to the next, a whole afternoon may be spent in examining five or ten bushes. As evening approaches, the ora prepares to lie up for the night among the bushes or, having made a successful kill, rests panting in the shade, before waddling out on to the savanna, with distended belly dragging on the ground, in search of a small water-hole in which it can cool off and drink deeply, and thence by a well-worn trail to its earth in a vine-covered thicket.

It is fortunate that the Komodo people have their taboos regarding deer and pig, and that they fear the ora as *Boaja davat*, "the land crocodile," for these varans are as vulnerable in their indifference to man as are the giant tortoises and iguanas. Both Attenborough and the Swedish photographer, Sven Gillsäter, found that they could photograph the giants among them with portrait lenses at ranges of less than 3 feet when they were feeding on bait, typically oblivious of yellow butterflies settling on their snouts. Although Auffenberg's researches have shown that the oras are more numerous than had been supposed, and their range more extensive, the latter is very limited. Moreover, if he is correct in estimating their total population at around 5,000, less than one-third of these can be mature adults, including only about 350 breeding females. The human population explosion in Indonesia must inevitably threaten the continued existence of any animal with such a restricted breeding potential and habitat.

10: Turtle Mysteries

At high tide on spring or summer nights, when the horizon is blurred at twilight or in the pale blue moonlight, shadowy forms are borne by the rollers to the beaches of the Galápagos and the Seychelles and many another remote island or atoll or tropical shore. In the last wave a head is seen, peering in every direction, for even the light of a match struck far up the beach may frighten the shining box-like creature—no Aphrodite—back to the ocean, through which it has travelled so far to this rendezvous. For some minutes it remains in the zone of breakers, rising and falling, with the water streaming off the rough texture of its shell. Then, with a few laboured struggles of its flippers, and with a snorting groan clearly heard above the booming breakers and the whistle of the wind through the mangroves at the back of the beach, the colossus frees itself from the surfing water and comes ashore where, in the Swedish photographer Jan Linblad's words, the "sea foam sensuously licks the smooth land." A turtle has returned to her nesting beach.

There are a number of varieties of turtle: green turtles; flatbacks with less domed carapaces and larger heads; small hawksbills less than 3 feet long, with yellow plastrons and brown carapaces whose scales overlap, in contrast with the smooth shells of green turtles; reddish or brown loggerheads with distinctively broad heads, a flipper span of as much as 9 feet and weights of up to 850 pounds; ridleys, smallest of all, with an average length of less than 2 feet (though the breadth of the carapace may exceed this by several inches) in comparison with the 6 or 8 feet of the gigantic leatherbacks, females of which weigh upwards of 1,500 pounds and males possibly 1 ton. To accommodate this great weight the leatherback differs structurally from other turtles, whose carapaces are covered with a horny layer of plates forming a symmetrical pattern. The leatherback's carapace is an immense shield of thick, tough leathery skin, saturated with oil, stretched over a mosaic of bony discs and seven longitudinal ridges or struts, unconnected to the skeleton, which act as a support like the ribs of a boat, as do the five struts on the plastron.

Some turtles, like the leatherbacks, the most pelagic, return from hunting fish and squid in the open ocean; others, like the hawksbills, from feeding on crabs and mollusks in coastal waters or, like the green turtles, from their vast grazing meadows of turtle-grass (*Thalassia*) and other marine grasses hundreds of miles in extent. The turtles are the only large animals to harvest these salt-water fields, the most productive of which lie under a few feet of water among reefs or archipelagoes or, as at Great Inagua, almost surrounded by mangrove swamps. The majority have travelled hundreds or even thousands of miles to rendezvous at the nesting beaches. We know very little about their oceanic wanderings, but during the months following the south-east monsoon, green turtles stream north from the vast *Thalassia* beds in the Mozambique Channel to their nesting beaches on Aldabra and other islands of the Seychelles, and also to remote beaches on the coasts of northern Kenya and Somaliland, while, as those who have read the present author's *The Unknown Ocean* will know, other green turtles migrate a minimum of 1,300 miles against the South Equatorial Current from the *Thalassia* fields off the coast of Brazil to nesting beaches on Ascension Island, a mere speck in mid-Atlantic.

We asked in that book how a turtle could recall the position of, and navigate over hundreds of miles of open sea to, a

Leatherback turtle, fastest swimming of the marine turtles, pushes up furrow of sand as it heads into the sea

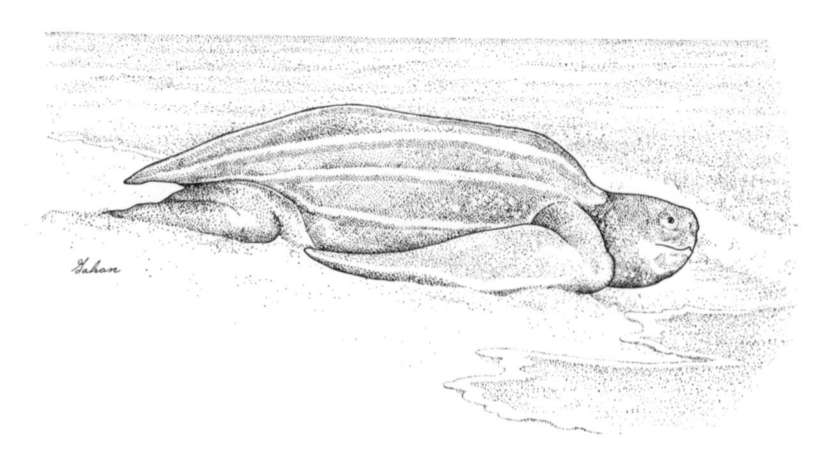

minute oceanic island. There was then no clue as to how such a homing feat might be accomplished, nor is there now. A recent suggestion that acoustics might be a factor involved, and that a homing turtle might pick up recognizable submarine echoes from an island, having first succeeded in navigating to its approximate locality, would not appear to be tenable; for turtles do not, so far as we know, make echo-ranging sounds as cetaceans do. However, frustrating as it is not to be able to provide a solution to this intriguing problem, the turtles continue to shuttle to and fro across the oceans between feeding grounds and laying beaches, and we do at least now know that 96 per cent of one sample of Sarawak green turtles have repeatedly returned to beaches previously patronized, and that the average Costa Rican green turtle actually returns to within seven-eighths of a mile of its customary nesting station on the Tortuguero beach, and that many do so to within a quarter of a mile, despite the fact that they have homed from feeding grounds dispersed over an arc of 2,000 miles from Yucatan and Cuba to Colombia. The urge to return to its birthplace or the place where it rears its young is dominant in every migratory animal, and this impelling force, rather than the limited number of beaches with suitably sandy nesting areas, perhaps accounts for heavy concentrations of turtles at certain beaches. The numbers of green turtles homing, for example, on the Aves Islands, which are barely exposed banks only 500 yards long and rising less than 8 or 10 feet at their highest point above the surrounding turbulent seas, are so great that late comers unavoidably dig up eggs laid by earlier arrivals, when excavating their own nests. Yet, while *en route* to these insignificant banks, 100 miles south-west of Guadeloupe, they bypass extensive apparently suitable nesting beaches. This would seem to be a wasteful departure from Nature's normally neat economy, and is perhaps explained by the fact that, because these islands have decreased in size since the last century, the turtles are continuing to return to a traditional nesting place, much of which has subsided beneath the sea.

A year or two ago such behaviour by green turtles would have been described as abnormal, for to congregate at one time in such numbers at their laying beaches might be considered a limiting factor to any increases in regional populations, and also render them peculiarly vulnerable to both natural and human predators; but we now know that on the island of Europa, in the Mozambique Channel, where it is estimated that the world's largest colony of some 8,000 green turtles nest along a 4-mile strip of beach, more than 700 females may be laying in a single night and digging up many nests excavated by those that had laid on previous nights. Moreover, the Atlantic race of ridley turtles also behave, or used to behave, in this manner. Until a film of ridleys nesting on a remote beach north of Tampico in 1948 was re-examined in 1961 virtually nothing was known of their life-history, for though solitary ridleys nested in colonies of other species on many Mexican beaches, the numbers doing so were patently insufficient to maintain a viable population. However, when this old film was critically reappraised it was realized that as many as 10,000 ridleys were nesting on a single section of the 90-mile-long beach at one time, and that upwards of 40,000 laid their eggs within the space of two or three days. Unfortunately, there is some confusion as to whether they normally nested on the Tampico beach in such densities that they were obliged to crawl over one another and inevitably, in such congested conditions, destroy the fresh nests of other turtles when excavating their own, as appeared to be the case in the film: or whether, as stated by the natives, they normally landed in three distinct groups at different places on the beach, and thus avoided this wasteful destruction. Since these immense gatherings, or *arribadas*, of ridleys are reported to occur, between April and June, on unpredictable dates from one year to the next, the thousands of individuals must converge from scattered feeding grounds hundreds of miles distant on a traditional rendezvous in the seas off Tampico, before at the appropriate mysterious hour being impelled by a mass impulse to go ashore, apparently in conditions of high winds and heavy surf.

That, prior to the rediscovery of the 1948 film, the existence of the ridleys' *arribadas* was known only to a few local inhabitants, was due partly to the remoteness of their nesting beaches, but mainly to the fact that because other species of turtles normally go ashore to nest at night, no one had searched the beaches for turtles laying by day. Indeed, it had always been assumed that turtles only came ashore under cover of darkness as a protective measure against predators, and that they never left the sea except for the sole purpose of laying their eggs. However, this assumption was not entirely warranted, for though they usually come ashore in overcast conditions of drizzling rain, they also do so in starlight and in the bright tropical moonlight. Moreover, odd clues are now coming to hand which indicate that night landings are not as universal as has been supposed, and it seems possible that centuries of persecution by the native inhabitants and by ships' crews may have been partly responsible for this nocturnalism. It is now known, for example, that at several remote Pacific beaches, such as those of the Galápagos, both male and female green turtles (or, rather, a black race of this species) haul out to bask in the sun on rock ledges and, incidentally, crop the leaves of mangroves bordering lagoons; while on the Pearl and Hermes reefs of the Leeward Islands (Hawaii) green turtles have been seen basking side by side with monk seals, the only tropical seals, whose world population of perhaps 1,500 is concentrated in the extreme northwest of the Leeward Archipelago. Quite recently, too, Cousteau has made a somewhat confused reference to watching some 2,500 turtles, presumably green, coming in with the tide, one behind the other, in blazing sunshine to lay on the broad sandy beaches of Europa, though he adds that laying continued throughout the night, with the return to sea beginning at first morning light. However, there are obviously grave disadvantages associated with diurnal laying, for Cousteau goes on to describe how these Europa turtles were not only plagued by a most virulent scourge of mosquitoes which covered their heads and flippers, but were also vulnerable to excessive exposure to the heat of the sun; for since every turtle invariably persisted in ploughing over the beach directly towards its objective, and would not make any attempt to go round an obstacle, a small rock or protruding root was sufficient to halt further progress, and the unfortunate animal soon succumbed to the sun—an astounding illustration of the limitations of a superb instinct capable of guiding a turtle across hundreds of miles of ocean.

The thousands of ridleys of the Pacific race lay singly by night along some 2,000 miles of beaches from Ecuador to Baja California; so what do those of the Atlantic race gain by laying diurnally and *en masse*? A suggestion that, with so many eggs laid in so short a period, the turtles' numerous predators are unable to dig up as great a percentage of nests as they are able to do in the case of small numbers of turtles laying at intervals, hardly seems sufficient explanation. The solution lies no doubt in some past experience. A green turtle, for instance, goes ashore six or seven times at intervals of a couple of weeks to lay batches of up to 200 eggs, 100 on average; but this she does only once in two or three or in some instances four years. Throughout this laying season of three months or more the females lie offshore. Although the males are also present at this time, and are indeed the first to arrive off the beaches to cruise around feeding on turtle-grass if available while surfacing frequently to look for newly arrived females, they never go ashore, except involuntarily when a breaker casts up a pair in the act of mating. Males seem to outnumber the females, and for a week or more

Mating green sea turtles

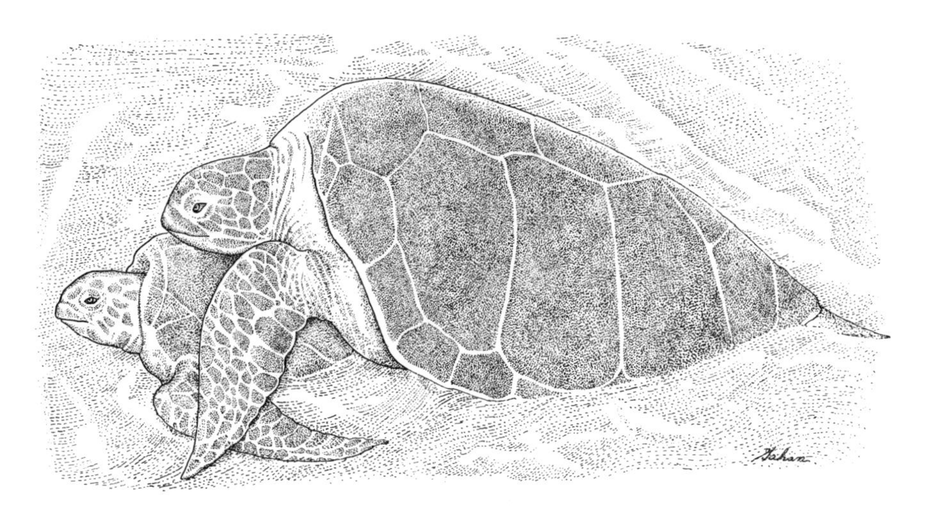

early in the season, when many mating groups are to be seen, a pair may be ringed round with as many as six or ten lusting males. Some females, however, do not mate until after they have returned to sea from the laying beach, and the males will therefore fertilize eggs that will not be laid until some years later.

Travis, when diving among the Seychelles, was struck by the large numbers of green turtles that had had one or more flippers bitten off, though even when minus one front flipper and both hind ones they were still able to swim well. He attributed this mutilation to sharks, which are partial to turtle meat, and describes in *Beyond the Reefs* how:

> Towards the end of each day the turtles would hide themselves away under ledges of coral or shelves of rock, within cave or crevice, in an effort to find safety from the night-prowling Tigers, Hammerheads and other such marauders. Only at full moon would the turtles continue with their love-making throughout the night and then only within the comparative safety of shallow water. Even this did not always provide the necessary security. . . . One evening Raville and Ton Pierre . . . borrowed my electric lantern and set off in a dinghy . . . to see if they could find any turtles at rest in the shallow water that lay between the fringing reef and the shore. . . . In the beam of light they saw a turtle, half in and half out of the water, struggling desperately to free itself from . . . something whose back was awash and whose large dorsal fin rose three or more feet into the air! At the same instant the turtle tore free and scrambled up the sandy ramp that led up the beach. It was almost clear when the Tiger Shark surged up after it. . . . Such was the fury of the brute that by lashing the shallows into foam with its powerful tail it succeeded in driving its heavy bulk right out of the water, up on to the ridge of the sand, where the weight of its body was supported by the two pectoral

fins splayed out like stunted arms on either side of its gross torso.

But let us follow the nesting female. Heaving herself forward and upwards out of the breaking waves with compulsive movements of her powerful flippers, mucus streaming from her eyes as glands excrete excess salt, she struggles ashore and begins to drag herself up the slope of the beach, thrusting with her hind flippers. In the water she is a strong swimmer, but on land, in Linblad's words, she becomes like a gigantic rock, like an astronaut on Jupiter weighed down by the force of gravity; for although the average female weighs only 250 pounds, those from Ascension Island and the Seychelles have been known to scale 700 or 800 pounds. Breathing presents a problem when she is out of the water, and she must rest frequently. After only 2 or 3 yards of laborious progress she raises her head, the bulging pocket of skin beneath her chin empties, and her huge lungs fill with another breath. Then, with a groan she sinks down to rest her head on the sand and lie there motionless for a long time. In this manner she may take as long as half an hour to plod up the typically steep slope from the sea to the nesting area. This is composed of lightweight sand that does not pack into a hard surface, and is situated high up the beach on a platform, free from such obstructions as drift-logs and the roots of coconut palms, above the level of spring-tides. She may then wander around for a further quarter of an hour, or perhaps for as long as three hours, before she ultimately finds a spot that suits her near a sharp rise in the beach, or at the edge of the most seaward vegetation. But nevertheless, the slightest disturbance, perhaps no more than the sand tremors resulting from other females excavating nest-holes, may cause her to abandon her operations and return to the sea. However, once she has begun digging, then nothing short of physical injury can prevent her from scooping out a pit in the sand from 8 inches to 3 feet deep and from 12 to 18 inches in diameter, though she must perforce stop to breathe heavily with deep sighs every 20 to 45 seconds. She digs with her 3-foot-long hind flippers, which are more suited to this operation than the front pair. In Linblad's words again, she operates like a gigantic computerized machine. The flexible tip of one hind paddle sinks down into the sand behind the tail and begins to excavate. After half a dozen digging movements a small portion of the sand held in the paddle scoop is raised, while at the same time the other paddle is being loaded. When this one is full the upper paddle jerks its load backwards and, as it goes down again, sweeps the sand aside to prevent it from sliding back into the hole. Every movement is perfectly co-ordinated for the production of a nesting hole in the most economical manner, and after an hour or so she completes the operation with a narrower shaft in the moist deeper layer of beach. According to the size of the turtle, so this factor determines the varying depths of the nest-holes which, as Carr has pointed out, are not in fact formless pits but elegantly flask-shaped, though slightly lop-sided, spherical chambers that communicate with the surface by a narrow neck.

Having concluded her excavations she rests again, filling her lungs with long deep breaths, before the first egg is released from the cloaca, and is followed by a succession at the rate of four or five a minute initially, but subsequently at twice this rate, until a pile of shining white eggs the size of billiard balls lie in the bottom of the chamber, bound together with a secretion that also cements its walls. There follows another breathing space and, as she rests motionless, the sac under her chin is filled with air from time to time and expelled with a groan; and she blinks regularly, squeezing out long viscous tear-drops. But finally she sets about covering the eggs to a depth of 5 or 8 inches with the moist sand thrown up in the course of her excavations, pushing this into the hole with sweeping motions of her flippers for ten minutes or so, then trampling it down. Then she sweeps and scuffles the sand all round the nest by "swimming" through the sand and flipping it back and forth for a further quarter of an hour, or even for a full hour. Though this is obviously a protective measure designed to conceal any trace of her activities, it is actually effective in preserving only a minority of the nests from such predators as coatis, foxes, feral dogs or monitor lizards, while even tigers and especially jaguars are partial to turtle eggs. Although these and other predators are apparently only able to scent out turtles' nests for the first 24 or 48 hours after the eggs have been laid in them, and monitors can be watched excavating site after site on a beach without locating a single nest, very large numbers of nests are rifled and raccoons have been known to dig up 70 per cent of loggerhead nests on a single beach. Both tigers and jaguars also prey on the females on the nesting beaches, first turning them over on to their backs, a feat that presents no difficulty to a tiger even when the turtle in question is a 1,500-pound leatherback. There is a story of one Malay egg-gatherer who, when sleeping at dawn beside the dying embers of his fire, was nipped by a young tiger, curious as to the nature of the prone figure enwrapped in a sarong as a protection against the sand-flies.

Her task concluded with so much mechanical labour, the turtle returns down the beach considerably faster than she had come up it though, even so, at a rate of only 6 to 12 feet a minute. This return is by a different route, and she leaves in her wake a broad trail resembling the tracks of a small tractor or amphibious tank, with a dotted line in the centre imprinted by her short tail which dips in the sand after each shove by her flippers; for both green turtles and leatherbacks, being the most highly adapted to marine life, progress on shore with the breast-stroke action they employ when swimming, whereas other species walk in the normal manner by moving their flippers on either side alternately.

Fifty to sixty days after the female has returned to the sea, though not before she has completed her laying programme for the season, the first batch of young turtles hatch out—perhaps less than half the full clutch. But four or five days elapse before they succeed in struggling to the surface, because they have considerable difficulty in breaking through the cap of sand with the horny projections on the tips of their snouts. Carr, who has made turtles his life study, has been able to reveal experimentally the fascinating details of what actually happens to the community of young turtles within a nest, and he describes how:

> The first young that hatch do not start digging at once but lie still till some of their nest mates are free of the egg . . . turtles of the top layer scratch down the ceiling. Those around the sides cut down the walls. Those on the bottom . . . trample and compact the sand that filters down from above, and they serve as a sort of super-organism, stirring it out of recurrent spells of lassitude. Lying passively for a time under the weight of its fellows, one of them will suddenly burst into a spasm of squirming that triggers a new pandemic of work in the mass.

Even when near the surface a community of young turtles does not break out immediately, but waits for some stimulus, possibly associated with temperature or humidity; and when they do emerge during the course of a couple of hours, it is usually after midnight and after a drizzle of rain. Nor do all the young from one nest break out on the same night, but at intervals over a number of days; perhaps it would be more correct to state that this has proved to be the case among the green turtles of Sarawak, for one would have supposed that a partial break-out would leave those young still in the nest exposed to predators. For that matter, should a herd of waris, the South American white-lipped peccaries, happen by ill-chance to visit a beach when the young turtles are still lying in their nests just beneath the surface awaiting the "signal" to explode, they will clean up the entire beach. However, once the young are above ground there is no further delaying, and though their carapaces are still soft and the lids hooding their eyes are only now lifted, they instantly scuttle off in the direction of the sea on what amounts to a death run. Not only have they to negotiate what is for them—barely as large as the palm of one's hand—the tremendous obstacle-course presented by the pitted surface of the beach, but they must also evade the numerous predators which, by some mysterious agency, appear to be aware of the exact hour when thousands of young turtles will be making their exodus. True that, as in the nest the activity of one little turtle served to overcome recurring bouts of lassitude among its fellows, so now on the beach one encourages another to renew its efforts to reach the sea; but nevertheless, according to locality, crabs, gulls, night-herons, opossums, dogs, and coyotes gobble up the majority while they are still on the beach. Today, on less remote beaches, the young turtles have also to compete with a new hazard, the headlights of cars, for though the simplest explanation of their ability to orientate towards the sea is that they follow the downward slope of the beach, they also apparently orientate towards the lightest horizon, which even at night lies over the sea, and which they can detect in all conditions except heavy rain. However, artificial lights may offer a confusing choice of horizons, and it has been claimed they can be led in any direction by the beam from a torch.

If the majority of young tortoises which exodus under cover of darkness have to evade so many hazards before they can reach the sea, what must be the fate of the minority breaking out during the daytime and subjected to the additional hazard of being shrivelled by the sun after a few minutes' exposure? If it is true that all young ridleys exodus during the day it is difficult to understand how any can survive, though it has been suggested that with so many tens of thousands breaking out at one time, predators are unable to cope with their numbers, with the result that an abnormal percentage reaches the sea. But it is significant that though the majority of the young green turtles on Europa broke out during the day, all in Cousteau's experiences were instantly gobbled up by the "crows" and frigate-birds breeding on the island, and only the few that emerged at night survived. Cousteau, incidentally, also refers to a unique state of affairs in the Red Sea where, on the desert island of Mar Mar on the Yemen coast, he found a vicious and continuous state of war being waged between boobies, nesting close to the sea, and turtles which went ashore by night to devour their nestlings. These were presumably flatbacks, which are known to nest on an island in the Gulf of Aden, where they are extensively preyed on by feral dogs. Although only green turtles are vegetarians, and even they will feed voraciously on the shell-less "flying snails" and on small lobsters, there are clearly strange happenings in the Indian Ocean which Archie Carr ought to investigate.

Even when the few young turtles that survive the crossing of the beaches reach the sea they have still not gained sanctuary, for a multitude of large and small marine predators await them, and experiments suggest that for the first forty-eight hours they paddle non-stop with steady strokes of their relatively long flippers away from these dangerous shallows, swimming constantly against the flow of the water until eventually picked up by an ocean current and carried away to those nurseries in which they can begin the long period of growth to the massive proportions of their parents. The localities of these nurseries, assuming that young turtles do in fact concentrate in special feeding grounds, are still not known. No one has ever caught or seen yearling turtles in the open sea, despite the fact that they must be somewhere near the surface where food in the form of plankton or algae is present. Carr has suggested that floating fields of sargassum weed may provide nurseries for the young of some species, and there is some evidence in support of this possibility, for young green turtles hatched in captivity eat sargassum weed, sleep supported on rafts of it, and explore among it for small soft-bodied animalcules. In addition, there is a record of nine young loggerheads being captured on a raft of this weed in the Gulf Stream off Florida; while when large quantities of sargassum weed were being washed up on many Florida beaches in October 1970, a mere twenty minutes' search through the mats of weed on one beach, on a dark, rainy, windy night, resulted in the capture of another five young loggerheads. It is possibly significant that their reddish or brown colouring camouflages them among the brown and yellow fronds of the sargassum.

11: Mud-Skippers

We have seen that mangroves, by fixing the soil on sand-cay or atoll, play an important role as builders of islands and nurturers of forests; but their evergreen jungles of small and medium-sized trees, with canopies of glossy light-green leathery leaves, growing in the zone between mean sea-level and the high water of spring-tides on sheltered ocean shores and in river estuaries where salt water intrudes, also provide a habitat for a varied and numerous swamp fauna. The latter ranges from lordly tigers and jaguars down to humble shore snails, barnacles and oysters which cling in clusters to the mangroves' prop-roots and close their shells in unison with the clatter of an aero-engine when the tide ebbs back. Moreover, the salt tidal water, being denser than the fresh river water, flows beneath the latter, making it possible for marine fauna to travel upstream into the swamp channels. Sea fish such as gray mullet and flounders prey on the estuarine fauna, and others feed on organic waste carried downstream from swamps and salt marshes. The fact that the salt tidal water travels up the estuary in the form of a wedge, creeping along the bottom beneath the fresh water and tapering upstream, accounts for the ability of mangroves to grow far inland, providing that the salt-water wedge is not impeded by sand-bars and that the banks of the estuary are not too high for the roots to reach down to the wedge on the bottom mud. In favourable conditions, such as those prevailing in the Sunderbans in the Bay of Bengal, mangroves can extend up the Hooghly River for scores of miles beyond Calcutta.

The mangrove swamp, like the estuary, is the meeting place of the fauna of the land and of the ocean, of fresh water and salt water. The mixing of these waters obliges some of its inhabitants to tolerate a more saline environment than is normal for their kind, while others must adapt to a less saline one. Some, like the four-eyed fish of Central and South American estuaries, bypass the adverse effects of salinity (and also of strong currents and silting) on their eggs by bearing live young; while both the adults and tadpoles of a crab-eating frog in Indonesian swamps are tolerant of brackish water. The distribution of the fauna in this inter-tidal zone is therefore determined by the degree of salinity, the period of exposure between tides, and the composition of the mud or sand. There are, as we have already noted, no limits to the expansion of mangrove trees, and in time their dense, almost impenetrable labyrinthine thickets and jungles—hot, steamy and gloomy—become the central area of every swamp, and the only practicable way to traverse the maze of stilt-roots curving out from the bases of the trees and hooping down to the mud several feet away, is by stepping warily from hoop to hoop. Trapping, with their peculiar root systems and pneumatophores, silt and other organic matter, the mangroves consolidate this and in course of time a rich grayish-black mud reaching a depth of several yards is composted. At high tide the mudbanks are covered to a depth of 3 feet by turbid black water; but when these dry out at low tide except for the network of runnels and channels, they swarm with crabs, including fiddlers and robbers. The higher above sea-level a particular area of swamp, the less frequent the flooding and the firmer the mud; and each tier is the habitat of different plants and animals. One kind of mud-skipper, for example, inhabits the zone of semi-fluid mud, but is replaced on less liquid mud by a second kind, and by a third on firmer mud. On slightly higher ground small mud-burrowing crabs predominate, but are replaced on more clayey mud by larger crabs of the same genus which live in the holes of the mud-lobsters.

There are so many little-studied species of the tropical mud-skippers swarming over every mudbank and mangrove swamp that one can only generalize about their habits. To human eyes these small fish, from 4 to 10 inches long, are exceedingly comical and endearing, both in their actions and in their appearance, for their slim brown bodies with blue or orange speckles have disproportionately large heads with relatively gigantic square mouths, while their prominent bulbous eyes, resembling those of frogs rather than fish, are placed, almost touching, on the raised crowns of their heads. No doubt this periscopic mounting, enabling them to detect large objects at ranges of as much as 30 feet, and also the extensive rotary movement of which the eyes are capable, is a very necessary protective measure against the numerous birds alert to pick them off the mud, for the skippers are as much at home, perhaps more so, out of the water as in.

Although they can swim under water like any other fish, and remain submerged for at least ten minutes, they more commonly swim at the surface with eyes protruding or heads held erect above water. While doing so they frequently spring up, either to hydroplane, propelled by lashing tails, with head and body clear of the water, or, by curling their bodies and then straightening them with a rapid flicking motion, ricochet over pool or channel in a series of 2 to 4 foot bounds. But their pectoral fins, employed as stabilizers by normal fish, have been modified into sturdy, fleshy, jointed limbs with which they can haul themselves with a sculling motion out of the pools around the mangrove roots. Once on the flat they "row" themselves over the soft mud by swinging forward on their pectoral fins, employed as crutches, and then transferring their weight to the pelvic fins, leaving behind them a characteristic herring-bone trail; or skip and spring over the mud in 1 to 2 foot bounds. Jumping this way and that, they are much too agile to be caught before they take refuge in a crab's hole.

As the tide comes in, so the skippers migrate up the channels among the mangroves and disperse over the high-tide mud, where among the 6 to 12 inches high pneumatophores there may be as many as 1 to every 10 or 15 feet. Most feed on minute sandhoppers, crabs and other crustaceans. Some are herbivorous, skimming off a thin layer of mud and algae from the surface of the water with side-to-side movements of their heads; or, if at the sea's edge, nibbling at the seaweed on the smooth rounded boulders, from which the waves, breaking in full force, cannot dislodge them, for when the crest of a curling wave is about to crash on top of them, all the skippers turn to face it at the last fraction of a second, while clinging to the weed with the "suction cups" formed by their fused pelvic fins. Others climb up into the mangroves in hundreds just ahead of the flowing tide to perch on the arching roots and strike at flies and mosquitoes with their multi-purpose pectoral fins, with which they also clean themselves.

The problem is, of course, how does a fish with gills breathe when out of the water? Under water the skippers breathe like normal fish, drawing in water through the mouth and passing it out through the gills; and since, when perched on mangrove roots or rocks or at the edge of a runnel, they often do so with the tips of their tails dangling in the water, it was formerly believed that they were able to breathe by absorbing oxygen through their tails which, being richly supplied with blood-vessels, would act as supplementary gills. However, there does not seem to be any satisfactory evidence that this is the case, and it is now known that they can retain supplies of oxygen in pouch-like gill chambers between mouth and pharynx. But is this oxygen obtained from the air or from the water? Robert Stebbins and Margaret Kalk investigated in some detail the life-history of one species of skipper, *Periophthalmus sobrinus* on the Ilha da Inhaca 20 miles off Mozambique. This is one of the world's most southerly coral reefs because of a fortuitous combination of favourable currents and sea temperatures, associated with an absence of silting on the sheltered side of the reef. Even so, large portions of the reef are periodically destroyed by mud and debris swept out into the Bay of Lourenço Marques by river floods, and storms subsequently break off great blocks of the killed coral and cast them up on to the inter-tidal mudflats, exposed at low water, where they provide habitats for many kinds of primitive marine creatures on their shaded under-surfaces. Stebbins and Kalk observed that when an Inhaca skipper emerged from the water on to a mudbank it usually paused at the water's edge and gulped air which, together with water, was retained in its gill-chambers and provided a source of oxygen while it was ashore. Moreover, when a feeding skipper made a sudden lunge to seize its prey, it simultaneously expelled air and water through the gills with such force that the amusingly audible belch could be heard at distances of 10 to 20 feet, while wet patches or bubbles appeared on the surface of the partially dried mud. It would then immediately crawl to the nearest pool or tidal channel, replenish the water in its gill-chamber by means of a few seconds' rapid pumping with mouth under water, and once again gulp air at the surface.

On the other hand we have already noted that the skippers commonly swim with heads above water, as if for the purpose of breathing, while it could be argued that since muddy tropical waters are poorly aerated, because the bacteria in the decaying vegetation exhaust the supplies of oxygen, it would not be possible for the skippers to obtain oxygen from swamp pools and channels. Other naturalists believe that a skipper breathes through its skin, absorbing oxygen in much the same way as a frog does, and that if its skin dries out it suffocates. Certainly skippers never venture more than a few yards from the nearest pool or channel, and also moisten their skins by rolling over on wet mud; but on the other hand they can be seen sunbathing while lying on sides or backs and waving their fins to ward off mosquitoes, which emerge from their breeding places in crab burrows to bite their heads and backs.

The skippers are never more comical than when mating, and Attenborough has described how on a Madagascar beach, where droves of skippers had clambered on to the boulders near the tide-line, a large male resting on one boulder, blinking and retracting its eyeballs into their sockets, was joined

Mud-skipper male displaying for his mate on mudbank

by a smaller skipper. After lying side by side for a minute or so, the male hoisted its long rectangular, yellow and peacock-blue rear dorsal fin, and followed this by flashing up the gorgeous forward dorsal fin, while at the same time tremulously arching its body towards the female. To this display she responded by erecting her fins, and the two continued to flirt in this manner for the next ten minutes.

When feeding, the males stake out private territories, the maintenance of which involves considerable chasing of one individual by another. During the breeding season the male also establishes a nest territory, and there may be twenty nests, 4 feet apart on an average, on the exposed mud just below the seaward edge of the mangroves. In his territory the male excavates a Y-shaped burrow, the two entry arms of which link up from 2 to 4 inches below the surface and continue on down as a single tunnel to the brood chamber at a depth of 10 or 12 inches. By carrying up pellets of mud in his mouth from the interior of the burrow and placing them around the two entrances, or by squirting the mud out in a viscous stream, as if ejecting toothpaste from a tube, two turrets or chimneys about 2 inches in diameter are raised from 1 to 3 inches above the surface, one usually slightly higher than the other. In due course the burrow is filled to the brim with water.

Having constructed a burrow, the male can now set out to attract a female by a variety of complex displays. Different species of skippers will certainly employ different attraction displays, but the male of one species, after raising himself slightly on his pectoral fins, erects his dorsal fins with their extended and flexible orange rays. Then, by repeated vigorous and sinuous movements of his tail and a concave arching of his body, with head and tail raised off the mud, he makes three jumps, the first of which is the highest, repeating this performance as often as necessary to attract the attention of a female. If one approaches, he raises himself again on his pectorals, and inflates his orange throat and blows out his gills until his already large head appears twice its normal size, before turning towards his burrow, with the female following close behind. He enters, she hesitates. He reappears and begins to blow bubbles. She enters or, alternatively, leaps to one side when he dashes at her with intent to drive her in; and his courtship must begin again.

Mud-skippers are extremely pugnacious, repeatedly leaping at or over each other like game-cocks. Once he has a mate safely installed in his burrow, the male jumps around outside, chasing away other males and also crabs, threatening them by raising and lowering his dorsal fins flashing their black, white and orange rays: or, in the case of one species, the interior of whose mouth is coloured dark indigo-blue, by pushing, mouth to mouth, against another male.

Some species apparently construct burrows at all seasons of the year, for during the winter those on Inhaca remain within their turrets until the morning sun shines directly on these. As fish, the mud-skippers display a most extraordinary tolerance of heat. During the summer indeed they are much less active on cool cloudy days, and Stebbins and Kalk noted that although they avoid water and mud whose temperature is higher than 95 degrees F, they probably approach the extreme among fishes in exposure to solar radiation when frequenting unshaded mudflats under the direct rays of the tropical noon sun, while the heat absorbed in such circumstances is exceeded perhaps only by those fish that live in hot springs. All in all, the skippers, by their ability to retain air and water in their gills, and to feed both in the water and on land, have taken the fullest advantage of their niche in the muddy saline waters of mangrove swamps, inimical to most fish.

The skippers' habit of swimming with their heads above water may be in order to search for small surface prey. The 8-inch-long archer-fish of Indian and Pacific mangrove swamps and especially estuaries also have remarkable visual powers, which enable them to shoot down insect prey as large as dragonflies from beneath the surface of a channel by squirting jets of water from their mouths accurately at targets as much as 3 to 5 feet distant. Even young archers, only 1¼ inches long, can spit to 8 inches with precision. When the tide flows in among the mangroves the archer-fish swim ashore and take up vantage points just below the surface. On spotting an insect, the archer swims as near to it as possible, focuses on it with both eyes, and positions itself at an oblique

The archer-fish can squirt water up to 15 feet

angle below the target. Then it takes water into its mouth and, by suddenly closing its gill-covers, expels the water through a form of compression barrel fashioned by placing its tongue against a groove in the roof of its mouth. At the instant of firing, the tip of the archer's upper jaw is thrust through the surface of the water, and the force of the jet is powerful enough to knock a cockroach off its perch, providing that the prey is in an upright position so that the jet strikes it from below; but if the target is not dislodged by the first shot, the archer follows it up with a succession of jets, one after another. On completing its firing programme it then immediately snaps voraciously at whatever falls into the water, including inedible objects.

Although the archer-fish also feed on small swimming and floating prey, they are so dependent on insect food that they travel up-river into irrigation ditches and even flooded rice paddies, where they compete with freshwater fish. It is somewhat idle to speculate as to how the archer-fish evolved this remarkable hunting technique, though it is true that many of those fish that hunt for food in the sand or among plants on the bottom expose their prey by blowing away the upper layer of sand or plant debris with jets of water. Porcupine fish are one kind that do so, and two of these fish which were accustomed to being fed at the surface of their tank in a laboratory, would, when anyone approached, swim up and spit water at random above the surface in their excitement at the expectation of food. Such a habit could conceivably be refined to its present perfection in archer-fish if a species of bottom-living fish was obliged to become a surface feeder, though it is impossible to imagine under what conditions such a change could be effected.

12: The Diverse Inhabitants of Mangrove Swamps

In most parts of the world the swarming mud and runnel fauna of the mangrove swamps provide food for an immense and varied population of predatory birds—storks, herons, egrets and ibises, cormorants and darters, kingfishers, fish eagles, ospreys, and kites; and mangrove jungles form ideal breeding habitats for herons, egrets and ibises. Linblad has described how in 1965 he watched 3,000 scarlet ibises and many more cattle egrets assembling to roost every night in a single mangrove, whose dense canopy spread over a circumference of some 70 yards in the Caroni Swamp on the island of Trinidad. A hundred years ago scarlet ibises were extremely numerous in the mangrove swamps and flooded llanos (plains) of Colombia, Venezuela, Guyana and northern Brazil; but with the introduction of automatic firearms and outboard motor-boats, their flocks have been decimated, not only for food but for the brilliant plumes which are much in demand for carnival costumes; and although there are perhaps still some 7,000 in the Caroni Swamp, they are in some danger of extinction. It is to be hoped that Linblad has not written their threnody in the following passage:

> In the late evening as hundreds and thousands of them cascade down to the marsh there is a murmuration of sound under the reddening skies. . . . The red ibises move to the trees where they will spend the night at the very time when the light is saturated with red from the setting sun. The clear, cloudless evening sky suddenly glows with red, the wind whistles through hundreds of feathers, as a flock comes in fast with powerful wing-beats to stop suddenly, as if at a word of command, and glide silently on, their outlines etched against the darkening sky. A muffled murmur of wings momentarily caresses the twilight before they flutter down into the soft green mangroves.
>
> The birds continue to come in closely packed, weighing down the branches. . . . The short, nasal calls blend with the sound of wings brushing against the foliage of the branches and making crackling sounds as the dry twigs break. The steady stream of sound is like the crackling of a bonfire. There is an overall impression of fire: the fluttering red shapes leap like flames to the accompaniment of the breaking of dry twigs . . . the last rays of the sun leave the highest treetops, the clouds above light up the scene with a gentle, diffuse but still quite strong light. The birds are now almost self-illuminating, as though they are carrying red sidelights. Two thousand ibises, perhaps more, now decorate the trees in front of us, making them look as though they were laden with red blooms. . . . The evening invasion of the trees became more and more spectacular as the foliage was drowned by a tornado of red wings. The branches . . . of the tree . . . shone like

> clusters of rowan berries. . . . Then, with a rush of wings, the whole enormous bouquet of roses explodes into the sky and fills it with thousands of fluttering wings. Rocking and swaying in a column above the tree the flock hangs in the sky. . . .
>
> In the morning a flow of blood-red ibises poured out in a seemingly endless flood over the water in the channels. The visual impact was tremendous . . . this living flame of fire . . . the red feather guise is all of a piece, like the scarlet cloak of a Roman general, artitistically uniform except for the four outermost wing primaries, which are dipped in black. The black areas are actually iridescent green . . . and the birds glow like freshly oxygenated blood when seen under the first slanting rays of the morning sun. As the ibises gather in open flocks over the dark water, it is as though they are burning with their own inner light.

It is interesting to contemplate, as a digression, that a few years before Linblad's expedition to Trinidad there would probably have been no cattle egrets (buff-backed herons) roosting on the mangroves in the Caroni Swamp, for until 1930 these egrets were known to breed only in the Old World —in southern Spain and Portugal and various parts of Africa and Asia. But in that year a small number migrated across the Atlantic from north Africa to Guyana, where they found conditions among the herds of Brahman cattle so familiar and congenial that some were encouraged to begin breeding. In this respect, it has been suggested that egrets following cattle obtain from 1¼ to 1½ times as much food, for an expenditure of only two-thirds as much energy, as when feeding where there are no cattle, because the latter flush out the insects and other small game on which egrets feed, as do men cutting sugar-cane or cultivating the cane-fields with tractors. In any event the Guyana nucleus thrived so well that during the next twenty years its members colonized Surinam, Venezuela and Colombia. Such a phenomenal dispersal suggested that this new American stock was being regularly reinforced by further immigrations from the Old World, and confirmation of this was provided by an egret which, ringed in 1936 in the Coto Doñana in south-west Spain, was recovered the following year in Trinidad after a voyage of 4,000 miles by the direct Atlantic crossing, but much longer than this by its probable route via west Africa and the Cape Verde Islands.

Actually, the cattle egrets' colonization of the New World may not have been so dramatically sudden as asserted, for there had been probable sight records in northern districts of South America between 1877 and 1882, in Guyana and Surinam in 1911 and 1912, and in Colombia in 1916 or 1917. Moreover, it now seems likely that this trans-Atlantic migration may have been in operation for many years, unnoticed. One report refers to frequent sightings of cattle egrets on St. Helena, while late in October 1952 one actually alighted on a trawler fishing the Grand Banks 200 miles off St. Johns, and vagrants reported from Newfoundland may well have been off-course trans-Atlantic immigrants rather than colonizers from the Caribbean nuclei. However, once established and regularly reinforced, the rapidity of their dispersal was dramatic. By 1954 there were 2,000 in Florida, by 1957 they were in Texas, by 1962 in the Galápagos, and by 1965 in Baja California. In addition to the Galápagos, more than 800 miles south-west of Colombia, they had also been reported at sea between Cocos and Claperton Islands, as if the mysterious westerly drive initiated in the Old World still retained its impetus after they reached the New World. When birds such as sand-grouse and collared doves embark upon these lemming-like eruptions these are usually triggered off by food shortages or by population explosions, but such crises have not, so far as I am aware, been observed in the egret colonies in southern Europe and north Africa. Moreover, the extraordinary thing is that a similar, easterly, dispersal was taking place during approximately the same period in the Indian Ocean and the Pacific. Cattle egrets were first reported on the Seychelles about 1915. In 1948 two large colonies were discovered in northern Australia, and since these were established breeding colonies, they must have been formed some years earlier by immigrants from Malaya. By 1963 there were cattle egrets in New Guinea and New Zealand, and the final dénouement would seem to be for egrets from east and west to meet in Polynesia.

Some predators in the mangrove swamps are so directly dependent on the specialized fauna of this habitat that they cannot exist without them. Fifty years ago there were probably hundreds of pairs of kites in the Florida Everglades. There, red, black, and white varieties of mangroves, growing to heights of 80 or 100 feet in the case of the red and extending up all the rivers as far as tidal water penetrates—where their growth is restricted to the dimensions of fringing hedges along every waterway—cover more than a million acres of the National Park and form what is perhaps the most extensive mangrove jungle in the world. The Everglades kites' staple food was a species of giant snail, *Pomacea caliginosa*, which they obtained by clambering about the reeds like bitterns and picking the snails out of their shells with their slender hooked beaks. Another inhabitant of the Everglades, that extraordinary bird the limpkin, which has been described as resembling a cross between a crane and a rail, is also totally dependent for food on the giant snails, and is equipped with a special nick in its beak with which to grip them.

That the very existence of a bird-of-prey should be dependent on an inexhaustible supply of snails would seem to be evolution carried to a dangerous absurdity, but Nature could hardly take into account the crassness of man. Modern drainage schemes have exterminated the giant snails in the Everglades, and the world population of Everglades kites has been reduced to perhaps a score of pairs at Lake Okeechobee (the big water) where the unique habitat of the Everglades may be said to begin, 100 miles north of the foot of Florida.

For that matter, a major ecological tragedy of the contemporary American scene is the dehydration of the Everglades. This vast wilderness of swamp and water, no more than 8 feet deep and choked with saw-grass—the Indians Pa-hay-o-kee (grassy water)—but 50 miles or more wide, has been termed the broadest river in the world, until it merges imperceptibly with the tidal waters of the Gulf of Mexico. But now there seems a possibility that the Everglades may become only an unique memory. A decade of droughts, together with shortsighted flood control measures and human pressures for piped water to Florida's east-coast playgrounds and space agencies have deprived the shallow Everglades of their essential replenishments of fresh water from Okeechobee's 700 square miles of sweet water 5 or 6 feet deep, with the result that they are drying out. The fish die in their thousands in the rapidly evaporating waters; increasing salinity is altering the vegetational complex, inhibiting small marine organisms from breeding, and thereby depriving the Everglades' special birds, reptiles and mammals of basic food sources. It may be that only the rerouting of adequate water reserves can save the Everglades.

Another denizen of the Everglades to have suffered grievously from their desiccation and also from the activities of poachers—50,000 were slaughtered in 1966 alone—has been the American alligator, whose numbers are reported to have been reduced from many hundreds of thousands to about 20,000; and an alligator, like every other inhabitant of a particular habitat, fulfils a role of ecological value. In contrast to crocodiles, which are restricted to the mangroves' salt-water or brackish zone, alligators inhabit the freshwater zone, usually hibernating in holes in river banks during the winter months; but whereas they can survive in waters with a temperature as low as 45 degrees F, crocodiles become helpless in such conditions and die. In the breeding season the female alligator scrapes together a 3-feet-high and 6-feet-wide nest-mound of mud mixed with moist debris—leaf mould, twigs, branches, shrubs—in a dry sunny place near water. This she shapes and smoothes by crawling over and around it, prior to scooping out a hollow in which she lays from twenty to seventy eggs in two tiers, covering them with material from the edge of the mound before levelling it off. The constant humidity, together with the fermentation of the nest material, generates a higher temperature within the nest than without. Neverthless, the female makes frequent visits to the nest during the period of incubation in order to moisten and re-shape it, and finally when the young hatch and she hears them croaking, to help them out. Some females, if not all, keep company with the young in pools near the nest for several weeks after they hatch, though they subsequently eat them if they can catch them. These pools are the celebrated 'gator holes which serve a vital function in the ecology of the Everglades as habitats for fish and feeding places for waterfowl, but which soon fill up if not kept open by alligators.

The estuarine crocodile, whose range stretches from the Sunderbans to the Solomon Islands, also makes use of an incubator, raking dead leaves and rushes into a mound as much as 27 feet in circumference a few score yards from the water's edge, and buries her similar complement of eggs 2 feet deep in it. She also excavates two wallows, wider than but not so long as her body, a yard away from the nest. These fill with mud and water, and in one or other of them she remains to guard the nest, and to maintain its humidity by splashing water over it with her tail, throughout the sixty days of incubation. Though frequenting mangrove swamps and travelling far up tidal rivers, the estuarine crocodiles are the most marine of their kind. They are indeed mainly fish-eaters, though for that matter fish is a substantial item in the diet of all crocodiles, and also the American alligator, until they become too large and sluggish to capture them. The estuarine crocodiles also prey on snakes, lizards, water-fowl and mammals, including the macaque monkeys, which feed on crabs and mussels in the mangrove swamps and cause considerable alarm among the mud-skippers. The monkeys they sweep into the water with their tails and, since large individuals reach lengths of 20 to 25 feet, and at one time probably exceeded 30 feet, they also attack man, and those in the Sunderbans have, like tigers, a particularly bad record as man-eaters.

All the other species of crocodiles excavate holes a foot or two deep in sand or soft earth near water and then, like turtles, cover their average clutch of fifty eggs with soil and smooth over the area around the nest. But though the females of some species remain on guard near the nest during the day, basking in the sun with gaping jaws, and Nile crocodiles indeed lie almost on top of their nests day and night, they are no more successful than turtles in preserving their eggs. During the three months or more of her long vigil, possibly without feeding, the female is likely to become torpid, with the result that monitor lizards in particular are able to dig up a considerable proportion of the nests, aided no doubt by the fact that several females may lay in close proximity. Maribou storks take advantage of nests uncovered by the monitors, and eggs are also taken by baboons, hyenas, warthogs and bush-pigs, while the tigers of Vietnam and Malaya are reported to be particularly partial to crocodile eggs. However, since adult crocodiles, unlike turtles, are virtually immune from predators except for man (who has exterminated them in many parts of their range), their numbers would build up very rapidly if they were not controlled at some stage in their development. Thus, in addition to high egg losses, there

Estuarine (or salt-water) crocodile

is also heavy mortality among the foot-long young crocodiles when they emerge a few minutes after the female, on hearing their squeaks and croaks, has uncovered the nest by wriggling and squirming around it until she has scooped out the crater down to at least the top layer of eggs—a very necessary act, since the soil has hardened over the nest. The young immediately head for the water and hide in the vegetation at its edge or, being very active, climb into bushes or trees and cling to reeds like chameleons; the female continues to guard them, chasing away predators with snapping jaws and swiping tail; but, according to locality, so monitors, small cats, maribou storks, herons, ibises, fish eagles, eagle owls, ravens, and ground hornbills all take their toll of them. Moreover, when they finally enter the water they become the prey of fish, water turtles, terrapins, otters, pelicans, and especially larger crocodiles which snap them up in the shallows as fast as they reach the water. Their one protection, according to C. A. W. Guggisberg, is that for several weeks or possibly months they can subsist solely on their large egg-yolks, and can therefore hide away and not expose themselves hunting for food. But relatively few, possibly no more than 1 per cent, survive the first year, though once they have reached a length of 6 feet they are more or less immune from attack.

Nature was at her most ingenious in evolving a crocodile, as Angus Bellairs has demonstrated in *The World of Reptiles.* Its ears are shielded by scaly flaps which can be raised to expose the ear-drums, though normally the flaps remain closed, except for a slit which opens when the crocodile's head is above water and presumably suffices for hearing. Its eyes and nostrils are positioned on the upper surface of head and snout, enabling it to see and breathe while the rest of its body is submerged, and the nostrils are opened and closed by special muscles which seal off the inside of the nose when the beast is under water for periods of as long as an hour. The nasal cavities are drawn out into long tubes which open just in front of the windpipe far back in the throat. Since this region can be sealed off from the rest of the mouth by fleshy flaps acting as valves, the crocodile can open its jaws under water, as it is obliged to do when drowning large prey, without inhaling water into the windpipe. Moreover, it can also breathe while dismembering its prey—which it does by rolling over and over with powerful strokes of its tail and literally twisting off limbs and chunks of flesh—providing that it is able to thrust its nostrils above the surface, for air passes straight back though its nose into the windpipe behind the valves, bypassing the mouth. Although the muscles closing the jaws are extremely powerful, enabling large crocodiles to drag horses and cattle and even immensely strong jaguars and tigers into the water, those opening the jaws are so relatively weak that a man can hold a crocodile's mouth shut with his hands, or so it is said.

Channels through mangrove jungles and coral reefs, estuaries and large rivers are the haunts of those marine mammals, the *Sirene*, comprising manatees and dugongs. They can be distinguished by their relative size—the former reaching lengths of 11 or 15 feet, the latter from 8 to 10 feet—and by their tails, that of the manatee being rounded, the dugong's forked. Moreover, dugongs, ranging from the Red Sea and the Indian Ocean eastwards to the Great Barrier Reef, live almost exclusively in the salt-water zone, though in coastal waters too shallow for sharks or killer whales; whereas the manatees of West Africa and the eastern seaboard of America from the Amazon north to Miami are tolerant of both fresh and salt water, though preferring the latter, and frequently travel up such large rivers as the Niger, and the Orinoco and the Amazon. The mangroves are almost their last refuge in the Everglades, and it is to the warm waters of the Crystal River in the west of Florida that they retreat when there is cold weather in the Gulf of Mexico, for they are reputed to be susceptible to pneumonia if the temperature of the water falls below 60 degrees F. In the latter conditions two snorkel divers, Roland Eichler and Helmut Albrecht, observed that the manatees passed much of their time sleeping on the bed of a warm spring, supported by their fore-limbs half doubled under their chests and with their muzzles buried in the sand. They can indeed remain submerged for as long as a quarter of an hour, communicating by loud squeaks in waters that are often turbid or muddy, though in normal weather conditions they bask at the surface of some overgrown creek for long periods, with domed snouts and nostrils above water, breathing every two or three seconds. If stranded ashore, however, their diaphragm muscles are unequal to the effort entailed in raising their 2,000-pound bodies at the intake of breath, and they soon suffocate. Those in the warm springs were usually surrounded by large numbers of fish, mainly snappers feeding on their faeces and small sun-bass occupied in cleansing their bodies of algae.

With the notable exception of some of the turtles, the *Sirene* are almost the only marine herbivores. Swimming leisurely through the water-weed, they sweep it towards their broad muzzles with their paddle-shaped flippers, or raise their heavy bodies partly out of the water to crop the vegetation bordering a channel, grasping it and plucking it with their mobile upper lips, which are divided into two halves. Manatees have no front teeth—only the male dugong is equipped with two tusk-like incisors—and their daily feed of 50 or 60 pounds of weed is masticated by horny plates in their palates and lower jaws, and ground up very finely before it can be digested and passed through 60 feet of intestines.

The global stocks of both manatees and dugongs have been drastically reduced, for though probably long-lived they are slow breeders, since the single young one is gestated for more than 150 days. Therefore, once the herds in one area have been decimated by excessive hunting, they take a very long time, even with protection, to build up their numbers again. Only skeleton herds of manatees remain in their former Amazon stronghold, though there are still many in the Belize coastal waters of British Honduras; while dugongs are now scarce in the Red Sea, almost extinct in the Gulf of Manaar (Ceylon) and, though more numerous in Australian waters, much reduced in numbers, especially in the Coral Sea where many are trapped in shark-nets.

Two hundred years ago there was a third member of the *Sirene*, up to 35 feet long and protected against the ice of the north-west Pacific's cold seas by a corrugated skin 2 inches thick. But within thirty years of its discovery in the shallow waters round the islands of Copper and Bering, Steller's sea-cow was believed to have been exterminated by sealers and whalers. However, in July 1962 a Russian whaler, hunting in mainly unexplored waters in the vicinity of Cape Navin, sighted six unfamiliar animals more than 20 feet long and with fringed tails and split upper lips; and it now appears that sightings of similar animals, possibly sea-cows, have in fact been reported several times in these shallow waters which do not freeze over in winter.

If the mangrove jungles of the Everglades are unique in the New World, those of the Sunderbans are unique in the Old World. Formerly an unbroken swamp of more than 6½ thousand square miles, stretching from the Tetulia river in the east to the Hooghly in the west, the Sunderbans form the deltas of the Ganga-Brahmaputra river system. Saline swamps and countless scattered islands still extend for 200 miles across the deltas' gigantic spider's web of intersecting rivers and creeks, some of which are mere ditches, and a few others broad deep channels which here and there open out into lake-like expanses. Sunderbans is a corruption of the Sanskrit *Sundaranana*, the beautiful forest; for the flat islands, only a few of which are as much as 10 miles in diameter, were at one time covered for the most part with dense mangrove and evergreen jungle, malarious and sparsely inhabited. But by the close of the nineteenth century almost half the jungle in Pakistan's 2,000 square miles sector had been felled, while the bulk of that in the Indian sector has now been sacrificed to human settlement. Result—an ever-increasing devastation by typhoons and tidal bores.

There is scarcely a dry foot of land in the mangrove swamps and "forests" of tree-like ferns, which are periodically flooded by high tides and flattened by tsunamis and typhoons. Two-fifths are under saline water, and in the remainder the water is brackish. Fresh water is only to be found at a depth of 4 or 5 feet on some sandy beaches and in rain pans in the jungle. To these few water-holes flock all the superabundant game; but during the dry season from November to May they, and the predator tigers, must often have to drink brackish water. For this curious habitat has attracted a rich and varied fauna, which formerly included the buffalo and the one-horned rhinoceros. Only forty years ago tigers were particularly numerous in the Sunderbans, sharing the islands with the rhino and buffalo and many species of deer, while the channels were infested with muggers—those snub-snouted

Indian marsh crocodiles which, like the long-snouted gharials, have been almost exterminated.

One might suppose a semi-aquatic habitat to be most unsuitable for tigers, though there are—or were before the jungles were shattered, poisoned and defoliated—also tigers in the swamps of Vietnam, where during floods an islet of 500 acres might have a dozen tigers hunting over it. But the Sunderbans formerly provided tigers with an abundance of their favourite prey, deer and wild pig, and with thick cover in which to shelter during the heat of the day and also during the cool nights at that period when the winters in the Sunderbans were apparently cold; while if water for drinking and bathing was often brackish, there was no lack of it. Their highways were the embankments between the rice paddies on the jungle edge, and they evidently learned to survive the perils of floods, for old reports refer to tigers living in trees for periods of two weeks or more in such circumstances, swimming out from their arboreal retreats to retrieve floating carcasses and capture turtles, small crocodiles and varanus monitors 8 or 10 feet long, though such small game as tortoises, frogs, scorpions, and beetles were also acceptable to a hungry tiger. Despite these environmental hazards, the density of tigers in the Sunderbans was perhaps greater than anywhere else in their range, and that they survived is indicated by the fact that the Sunderbans is the habitat of the large, richly coloured and thick-furred Royal Bengal tiger (so-called). Nevertheless, a proportion of these tigers seems always inexplicably to have experienced difficulty in obtaining sufficient natural prey, for from the earliest eighteenth-century records to the present day, the jungle-clad islands of the Sunderbans have borne a particularly evil reputation for man-eaters. But in other parts of their range tigers become man-eaters either because of a lack of natural prey or, failing this, of domestic cattle: or because physical disability or injury prevents them from hunting these, whereas man is by contrast the easiest of prey: or, in rare instances, because they have been brought up by man-eating parents. In the old days it was the salt-workers, plying their trade on long sand-spits projecting from the jungle in the Sunderbans, who were the chief victims of man-eaters, despite the presence of look-outs posted to watch for tigers; for the latter would stalk up to them along the many deep and narrow creeks, often overgrown with grass and thorn, leading from the jungle. And as lately as 1967–70 the mortality rate among the wood-cutters, the collectors of honey from wild bees, and the fishermen who are often marooned twice a day when the tides inundate the jungle, has averaged seventy-five a year. Today, the few score tigers remaining in the Sunderbans, and they were still there at the time of the 1971 cyclone and Indo-Pakistan conflict, have been generally assumed to be exclusively man-eaters, if for no other reason than that the surviving stocks of chital deer and wild boar are probably insufficient to keep them supplied. However, a recent field study, which can fairly be described as sensational, suggests that in fact about a quarter of the Sunderbans tigers never attack men, and that only about a third are invariable man-killers; and that moreover the incidence of attacks cannot be correlated either with the numbers of men working in the jungles or with the absence of natural prey, but is definitely linked with a high water-level and the degree of salinity in the water! While it is difficult to understand why these factors should be responsible for a tiger being a man-eater, it does offer a possible explanation of the long-term record of man-eating in the Sunderbans.

13: The Marismas and the Camargue

Thousands of years of see-sawing conflict between sea and rivers have created marvellous habitats for rich and varied faunas in the delta battlegrounds in many parts of the world. None are more attractive to the naturalist than the deltas of the Rio Guadalquivir in south-west Spain and of the Rhône in the south of France, both of which have been the breeding grounds of the most remarkable of all wading birds, the flamingo, whose life-history is unique.

For the ultimate 50 miles of its course to the Atlantic, the Guadalquivir flows through more than 600 square miles of marshes. These, the *marismas*, forming one of the largest marsh areas in Europe, are virtually cut off from the sea by an immense sandbar, the Arenas Gordon, which extends for some 40 miles from the mouth of the Guadalquivir to the Rio Tinto. Varying in width from 2½ to 7½ miles and rising to a height of 340 feet, its dunes equal in area the entire sand-hill system of England, and resemble a miniature Sahara. For four or five months every winter and spring in normal years the Guadalquivir floods the *marismas* to a depth of 1 or 2 feet with salt tidal water, a third more saline than that of the Mediterranean, though this salt flood is diluted by the heavy rains of October and November and by the fresh water stored in the chain of shrinking lakes and lagoons stretching across the Coto Doñana, the relatively high and dry country of some 67,000 acres lying between the sand barrier of the Arenas Gordon and the *marismas*.

The wonderful wildlife sanctuary of the Coto harbours about thirty kinds of mammals, a score of amphibians and reptiles, and some 140 species of breeding birds. Their habitats are the woods of stone pines and junipers among the dunes, whose unstable sands, perpetually whipped away by westerly winds from the Atlantic, advance slowly but inexorably inland, overwhelming even large woods of pines, 20 or 30 feet high, that have been weakened by extensive felling and by the operations of charcoal burners, who strip the trees of all but their topmost branches, depriving the wild boars of their huge cones. Beyond the dunes are sandy flats of open scrub, including rosemary and lavender and thickets of *Halimium* whose large yellow flowers, resembling those of the rock-rose, bloom profusely in the early morning but wither in the heat of the noonday sun; scattered park-like groves of ancient cork-oaks; and wide rides through dense scrub country of pistacio, gorse and broom, bramble and tree-heath up to 12 feet high. Moister lower-lying areas on the borders of the thick aromatic undergrowth among the trees and swamps at the edges of the *marismas* are the feeding grounds of several thousand wild boars, which grub up boggy hollows for their favourite delicacy, mole-crickets, and of course for roots, including strangely enough the enormous bulb of the luxuriant *Urginea maritima*, from which the rat-poison Squill is manufactured. A couple of thousand fallow and red deer also feed by the lagoons and on the borders of the *marismas*. Their fawns and calves are killed by vagrant wolves from the northern *sierras* and by the 150 to 200 resident Spanish lynxes, though the latter prey predominantly on rabbits and red-legged partridges.

During the winter the *marismas* resemble an inland sea, a land beyond the outer fringes of the known world, through whose shallow waters one can ride for 40 miles or more without ever setting foot on dry land, wrote Guy Mountfort in *Portrait of a Wilderness*; and over which geese, duck, and waders from all parts of Western Europe fill the sky with the rush of their wings, as flight after flight drops down to feed in the sheltered lagoons and to cover the mudflats. In the days of the legendary hunter-naturalist Abel Chapman, one might also have witnessed the most remarkable adaptation to a completely alien environment in the form of a herd of wild camels splashing through 3 feet of water in clouds of spray from islet to islet. Originally imported from the Canary Islands to the province of Cadiz in 1829, for use as pack and draught animals in transporting road-building materials and also to draw conveyances in some of the towns, they proved to be such a constant source of terror to horses on the roads, and such a nuisance in having to be kept apart from the horses in the yards, that their employment on the roads had to be terminated, though some were subsequently used for ploughing and other agricultural purposes until as late as 1869. However, a few years after their introduction to Spain, fifty or sixty of the original eighty were banished to the *marismas*. There they not only thrived in this semi-aquatic environment but also bred, producing young in February or earlier, despite the fact that throughout the Bedouin world every camel-master assists the male to perform the mating act. Moreover, they lived almost exclusively in the marshes, pasturing on the low islands or *vetas* and higher ground, though feeding on the sub-aquatic herbage in wet winters when, with no dry land visible from a distance they appeared to be living in a sea of waters, and only infrequently raiding the Coto Doñana to crop the tops of the young pines. There is no reason to suppose that they would not still be inhabitating the *marismas* had not poachers begun to slaughter them in the 1890s, reducing their numbers to a score or so by 1910, though it was not until 1950 that the last five were removed from the *marismas* to a compound.

During the winter floods the *marismas* are spattered with the small *vetas*, which are composed of marine shells and crowned with grasses, camomile, and thistles, and though varying in size from a few square yards to a score or two of acres, seldom rise more than a foot or two above the waters. But as the floods subside in April and May and the *marismas* are covered with herbage, so ground-nesting passerines, ducks and large colonies of waders and terns can begin to nest on the *vetas*, though every year the wild boars venture out of the scrub to gorge on their eggs and nestlings to the exclusion of almost all other food while these are available in abundance. But the spring growth withers quickly as temperatures rise to above 100 degrees F, and at the height of the four months' summer the water is drained from the marshes by the withering sun and the *marismas* become a scorched wilderness of almost uniformly dead-level flats—mile upon mile of bleached and cracked, salt-encrusted mud deposited by the Guadalquivir. The only living vegetation is the glasswort (*Salicornia*), which covers vast areas of higher levels with stunted and isolated bushes. Horse and cattle, distorted by the heat mirage, graze over the brown expanse of parched reeds and dry sedges which stretch to the indeterminate horizon. Around the few pools and lagoons remaining from the rapidly evaporating waters some of the *marismas'* 30,000 herons and egrets congregate to prey on marsh frogs and such fish as the plump crucian carp, which has the ability to bury itself in the soft mud at the bottom of a pool, if the latter dries up, and survive until the autumn rains, and also the tiny top-minnow, *Gambusia*, recently introduced to control mosquito larvae. *Gambusia* was perhaps an unwise choice, for in several other parts of the world its introduction has threatened the native species of fish, which in some places it has exterminated, while at the same time it has proved to be little or no more efficient at mosquito control than other less dangerous species.

Drought is a constant hazard to the breeding birds of the *marismas*, for if the autumn and winter rains fail and the *marismas* are destitute of fresh water, then the birds are deprived of both food and nesting shelter. Such birds as night-herons and squacco herons may not even attempt to breed, while little egrets may desert their nests, and the breeding population of others of their kind may be reduced by 90 per cent. Similarly, most of the ground-nesting birds on the *vetas* do not lay, and those that do, particularly on *vetas* lying within a couple of miles of the periphery of the *marismas*, are at the mercy of even more predators than normally, for in addition to perennial harassment by wild boars and marsh harriers, imperial eagles, black kites and ravens, they now have to contend with foxes and rats which establish themselves in rabbit burrows on the *vetas*, and with snakes which engorge both eggs and nestlings.

Little egret

If drought is an intermittent threat to the breeding birds of the *marismas,* a more permanent threat to some species is that sedimentation, coupled with the silting up of drainage channels, is raising the level of the marshes in this almost landlocked delta. This must result in diminished salinity in both soil and water and also limit the extent of the annual flooding by the Guadalquivir. These factors must in turn adversely affect such birds as pratincoles, which depend upon a certain salinity to restrict the growth of vegetation in their breeding areas on the *vetas*; and others, such as avocets and flamingoes which obtain their food in more saline waters, and can only establish breeding colonies when the salinity is adequate. Although a thousand flamingoes still frequent the *marismas*, compared to perhaps ten times that number at the end of the nineteenth century, none are known to have bred since 1941, and it is reasonable to suppose that their failure to do so can be attributed to the adverse effect of this fall in salinity. Whether or not this is the case, the permanent and highly saline lagoons of the Camargue in the Rhône delta have replaced the *marismas* as the flamingoes' main European breeding station.

Man has unwittingly retained the flamingoes' ideal level of salinity in the Camargue lagoons: firstly by a large-scale salt-extraction industry, involving the retention of brine-pans which evaporate sea-water pumped into a string of lagoons; and secondly by controlling the run-off of water from the large lakes or *étangs* by means of sluices in embankments along the seaward edge of the Camargue and also along the complex network of dykes across the salt-flats or *sansouire*. The 300 square miles of the Camargue, of which rather less than one-sixth are *étangs* of salt or fresh water, and rather more are marsh, lie within the two arms of the Rhône and were created by it, initially by boulders brought down when it was a very much larger and more turbulent river, and subsequently by ever-increasing deposits of fine sandy silt. Intermittent inroads by the sea have supplied the salt content, crystallized in the soil or in solution in ground and surface waters, whose numerous gradations from sweet to saline favour a wide variety of fauna and flora, and around which the major part of all life in the Camargue revolves. In lower-lying areas the salt content may be so high that, as on the banks of the Guadalquivir, virtually no vegetation can grow, and vast tracks of mudflats are bare except for sporadic clumps of glasswort; while the salt content of those *étangs* nearest the Mediterranean exceeds, as they dry out, that of the sea itself at high summer. However, the salinity is very much lower in the large central Étang de Vaccares, which throughout the dry summer receives large amounts of fresh Rhône water via the irrigation channels that flood the rice-paddies bordering the Camargue.

If the varying salt content is the decisive factor in controlling the Camargue's peculiar vegetation and fauna, these are limited both in variety and numbers by that curse of Provence, the mistral which blows for nearly half the year out of the mountains to the north-west. Indeed, except in June, July, and August, there are few completely calm days in the Camargue, and even during these summer months the wind may be very strong for periods of days at a time. The mistral is a highly destructive factor, not only drying out the soil and dehydrating the vegetation, but impeding the movement of all life on both land and water. Its only saving grace is that it grounds that other curse of the Camargue, the voracious mosquitoes.

The central and southern regions of the Camargue are dominated by *étangs*—extensive sheets of shallow waters, broken up here and there by sandy spits and islets, which are crowned by clumps or solitary "trees" of dark green tamarisk 20 feet high, whose weeping feathery sprays are ever in motion. The tamarisk is the only tree or bush that can exist on the *sansouire*, thriving at the edge of any *étang* or ditch that contains fresh water for much of the year, and maintaining a hold on life in soil with a slight salt content. According to the seasonal level of the floods, so the *étangs* are surrounded by no less vast and featureless, brine-whitened flats and smaller saline pans, baked and cracked by the heat of the sun into myriads of crinkly-edged segments or discs, hard enough to drive a car over, but turning with rain into a greasy slake over which the naturalist slithers one foot forward and two back. Immense expanses of the *sansouire* are covered with a dense low scrub of glasswort, the breeding grounds of pratincoles and stone curlew, and which to an artist might appear as Provence's counterpart to the heather moors of the Scottish Highlands. Seawards, there is an interminable waste of sandbanks and highly saline lagoons, frequented by avocets, oystercatchers, redshanks and black-winged stilts, shelduck, terns and gulls, with a mirage of inverted trees along the horizon of the Mediterranean, where the sun glints on breaking combers.

Bordering the *sansouire* and *étangs* are dense freshwater marshes or *marais* screened by luxuriant clumps of tamarisk, which almost conceal the nimble black cattle of the Camargue, belly deep in the scrub. Gigantic growths of thick-stemmed reeds and bullrushes tower 10 or 12 feet from the roadside dykes; and there are vast primeval reed-beds, where thousands of herons and egrets nest, and in which wild boars lair, as they also do in the groves of tamarisks. A narrow but dense strip of forest, predominantly white poplar and willow, alder, and ash, screens the Rhône.

Throughout the winter a thousand or more flamingoes remain on the *étangs* and *sansouire*, much of which are flooded; and the latter also provide winter quarters for thousands of duck, though the majority cannot stay to nest because so much of the water dries out in the summer. Wigeon are the most prominent, but there are also mallard and gadwall, teal, pintail, shoveller, and red-crested pochards, together with a few lesser black-backed gulls and Mediterranean herring gulls distinguished by their yellow legs and also by their deep voices, while little egrets fish in the muddy shallows with sudden darts forward and lightning stabs of sharp black beaks. A pink and black line across the blue sky, or a pale pink shimmering cloud, is a herd of flamingoes flighting from one *étang* to another or evading a plane droning in from the Mediterranean.

When the sun goes down on a winter's evening, and dusk falls quickly, it is bitterly cold in the Camargue. With the bone-piercing wind, and guns popping everywhere, one might be back on a Northumbrian slake, save for the marsh harriers quartering scrub and reed beds and flushing wisps of frightened snipe. And it is to the distant thunder of breakers, and the braying, honking, whooping, and sighing of the extraordinary flamingoes that the winter naturalist lamps his way to bed up the winding stone stairs of an old shooting lodge.

14: The Incredible Flamingo

Flamingoes are unique; unique in their habitat, unique in their food and method of obtaining it, unique in their breeding environment and habits, and unique in physique. So emaciated and attentuated do they appear in flight and so elongated are their extremities, with necks, cruciform wings, and legs outstretched, that they seem bodiless; but they are beautiful beyond compare when they spread their rose-scarlet and black wings. The geographical distribution of their world population, comprising half a dozen species totalling perhaps 6 million birds, is extraordinary, ranging as it does from oceanic atolls such as Aldabra and volcanic islands such as the Galápagos to the swamps and salt-pans of Caribbean islands; from the *marismas* and *étangs* of south-west Europe to the mud deserts, thousands of miles in extent, of the northern Kalahari, the soda-lakes of East Africa and the Ethiopian Rift Valley, and the corrosively bitter, milky waters of

Madagascar lakes; from the salt-lakes on the Turkmenian shores of the Caspian Sea, to Kazakh, the Kirghiz steppes and the salt-flats of the Rann of Cutch in north-west India; to lakes at altitudes of 7,000 feet in Afghanistan and 14,000 or 15,000 feet in the Andes, where in the volcanic regions of southern Peru and Bolivia and northern Chile thousands of square miles are surfaced with salt or borax, and the flamingoes breed on the highest lakes and saline lagoons, ranged round by volcanic cones. But these diverse habitats have one feature in common : all are saline.

The seasonal and perennial lives of all flamingoes are controlled by two factors—salinity and water-level. Both vary considerably from one year to the next at all their main stations according to the amounts of winter rain and subsequent flooding or drought, with the additional hazard in the Caribbean of inundation by tides driven by high winds or hurricanes. Not only does the flamingoes' food supply depend upon a correct balance of salinity and water-level, but so does their ability to breed in any particular year. The erratic alternation in breeding localities on the shores of the Caspian is, for example, attributed to increased salinity adversely affecting feeding waters. Moreover, although the level of salinity may be suitable for the organisms on which they feed, it may not be suitable for their breeding. There is, for instance, some evidence that the salinity of a feeding area can be at a much lower level than that of a breeding area. Herein lies the probable cause of the decline of breeding in the *marismas* and the corresponding increase in the Camargue.

The erratic migrations and distribution of the flamingoes can also be attributed to the unpredictability of these specialized requirements, which obliges them to be ever searching for suitable feeding and breeding localities. In 1962, for instance, they established a breeding station for the first time in South Africa's Cape Colony because of ideal water conditions on one lagoon. These requirements also explain their failure to breed annually and, for that matter, not to breed for a number of years in succession—a peculiarity that long puzzled naturalists, though this apparent irregularity has been due in part to the lack of observers in their inaccessible haunts. However, for flamingoes to breed every year is the exception rather than the rule. The Camargue herd nested in only sixteen of the thirty-four years from 1914 to 1947, and not at all from 1962 to 1968, while on the Great Rann of Cutch they may do so only once in three or four years, or possibly at even longer intervals for observation is as difficult in this remote region as in almost all the flamingoes' habitats. This being the case it seems surprising that they do not avail themselves more frequently of their ability to postpone the laying season for several months if conditions are initially unsuitable at the normal season. That they are able to do this has been proved both on the Rann of Cutch and in the African lakes. The 10,000 square miles of the Great Rann, and also of the little Rann—last refuge of a few hundred Indian wild asses—are composed of enormous quantities of fine silt deposited by a number of rivers. Though these flow from regions of low rainfall they are active during the July to October period of the south-west monsoon, when they totally immerse the flat waters of the Ranns, nowhere more than a couple of feet above sea-level, to a depth of a few inches or several feet. In addition, sea-water is blown over the Ranns by strong winds and from time to time the sea floods over them, impregnating the flats with salt. The enormous city of more than 200,000 nesting flamingoes, occupying one-twentieth of a square mile, is located on a flat dry islet, about

1 mile long and ½ mile wide, on the more or less sandy waste of the Great Rann and 6 or 7 miles from the edge of the hinterland. Since the breeding season coincides with the monsoon, the birds are able to build their nests in shallow water; but if the rains fail the Rann becomes a hard-baked desert of blistered clay and glittering blinding salt-crystals. In such conditions the flamingoes are unable to obtain either food or building material. On the other hand, if the monsoon rains are exceptionally heavy, the Rann is flooded to such a depth that, again, the birds are unable to nest. Rather similar conditions, though from different causes, prevail in the Caribbean where, on the Bahamas island of Andros, the flamingoes, or some thousands of them, nest among the mangroves and are liable to have their nests swamped by unseasonable rises in the water-level. In 1944, however, when exceptional flooding prevented the Great Rann herd from breeding, its members delayed nesting for five months and bred in unusually large numbers the following March, with the result that by April there were more than 100,000 occupied nests.

But even if the water-level is suitable for nest building the flamingoes cannot settle down to breeding unless food supplies are adequate to support their enormous aggregations, which may comprise a quarter or half a million individuals on the Etosha mud-pan in the Kalahari and as many as 3 million on Lake Natron in Tanganyika. So, before detailing the extraordinary sequence of events in the nesting cities, we must consider the flamingoes' unique feeding habits and the nature of their food, which varies remarkably not only seasonally and from one locality to another, but possibly within one restricted geographical area such as the Caribbean, where the flamingoes of Andros and Great Inagua feed mainly on the small molluskan cerithiums, whereas those on Bonaire Island off Venezuela apparently feed on brine-shrimps. What are a flamingoes' physical attributes? It has very long legs for wading in several feet of water if necessary: a very long neck for feeding under water in similar conditions: and a uniquely constructed beak that can deal with equal efficiency with algae and seeds on the surface, with diatoms and copepods, small fish and their spawn, insects and their larvae, and with small mollusks and crustaceans, annelid worms and even mud on the bottom. Moreover, they may be feeding in waters that are nearly saturated with salt or soda, and in conditions which, in the East African lakes in particular, would appear to be deadly to all life. Leslie Brown, who in 1954 discovered the colossal concentration of flamingoes on Lake Natron during an aerial survey, since their presence on a mud-flat 7 or 8 miles out in the lake is hidden from the shore by mirage, has stated that when African lakes shrink, they usually become more alkaline. Sodium salts (mainly sodium carbonate and its derivatives) become more concentrated until the water is so bitter that it is lethal for animals to drink, but its high carbonate content, together with the abundant sunlight and high temperatures on the floor of the Rift Valley, combine to make the water an ideal environment for diatoms and blue-green algae, which multiply in extraordinary densities until the lakes are coloured a rich pea-soup green. Nevertheless, the continual blooming of algae to support 2 or 3 million flamingoes over a period of months staggers belief. Brown has calculated that from the waters of the Rift Valley lakes the flamingoes extract between 2 and 5 tons of algae per acre per year, a yield comparing favourably with that from a first-class arable pasture.

Lake Natron and its smaller neighbour Lake Magadi, across the border in Kenya, are two of these alkaline lakes, mainly dry and composed of layers of crystalline sodium carbonate overlying foul black mud. The crystalline deposits are continually renewed by large perennial springs which create lagoons of strongly alkaline water. Since the lakes have no outlets, the water never escapes, the great heat results in intense evaporation, and the lagoons quickly solidify into layers of crystalline salts. The flamingoes feeding on the blue-green algae of these lakes must therefore do so in waters that are not only shallow, evil-smelling and at hot-bath temperatures, but three-quarters crystallized with a white or pinkish crust of soda, while leprous wine-red or coffee-coloured floes of this substance float on the waters and coalesce into a solid pink expanse which glitters in the intense heat, though the lakes are 1,000 feet above sea-level. At Lake Haddington in Kenya several hundred thousand flamingoes are obliged to stalk out of the lagoon on to its rocky banks in order to drink at the sources of hot springs and geysers whose temperature is 154 degrees F.

The flamingoes on the African lakes are predominantly of the lesser race, with deep-keeled beaks suitable to surface feeding, though a few tens of thousands of the greater race with shallow-keeled beaks for bottom feeding on insect larvae and small mollusks, associate with them. Those of the West Indies and the Caspian area are greater flamingoes, feeding mainly on brine-shrimps or on the inch-long spiral-shelled cerithiums, both very small and spread in a thin layer over the bottom. These, the flamingoes, thigh-deep in the water, stir up by treading or jigging with their feet while standing on their heads, as it were, with the upper mandible—triangular in shape and fitting closely into the thicker and sharply angled lower mandible—almost parallel with the bottom and sweeping rapidly back and forth like a vacuum cleaner. In this action the thick, fleshy, muscular tongue, which lies in a groove at the base of the lower mandible, though it is so large that it fills the whole cavity between upper and lower mandibles, works to and fro like the piston in a pump. Rapidly drawing water in and forcing it, together with mud and sand, out in little jets, it retains the incoming shrimps or cerithiums with fine flexible bristles or lamellae, arranged in lines on the inner surface of both mandibles, which lie flat as the water is sucked in and are erected as it is expelled. Brown suggests that the mass of food trapped by the lamellae may be rolled towards the tongue by the two mandibles working together like a pair of wool carders, while the tongue, which is spiked with backward-curving hooks nearly ½ inch long, automatically pulls the mass into the gullet. However, when watching at close range lesser flamingoes feeding on the microscopic growth of blue-green algae in the surface waters of Lake Natron, he was normally only able to detect the rapid pulsing flow of water being ejected as the lamellae filtered the algae from the upper inch or so of water. The lesser flamingoes prefer to feed in shallow waters, 12 or 18 inches deep, stalking slowly through them or standing in one place, while scooping in the algae by swinging their heads and tilting their beaks from side to side with a rhythmic scything movement. But they can also feed efficiently when swimming in deep open water, even when this has been whipped up into waves by the wind, despite the inconvenience in such conditions of continually having to move their heads up and down to avoid repeated submersion of their nostrils. Brown suggests that in these circumstances the lower mandible, being bulbous and composed of extremely light cellular bone, may act as a float, rising and falling with the waves and thus automatically adjusting the level of the beak in the water; moreover, their habit of feeding together in immense rafts smooths the waves, resulting in patches of relatively calm water.

But though the food and feeding conditions of the flamingoes in the African lakes are extraordinary, those in other parts of their range are even more bizarre. During the winter the greater flamingoes in the Camargue feed in quite an ordinary manner. Strung out in a pink and white line among the innumerable low islets across an *étang*, a herd of 250 stalk thigh-deep on their angular pink stilts very slowly through the shallow waters, stained a rosy pink with their reflections, from one feeding place to another. Plunging their long pink necks deep into the water, disclosing as they do so the blood-red gashes of their wing-coverts and brilliant rose-scarlet tail-coverts, they continually rake the bottom, trampling and uprooting the vegetation, including such algae as *Rupia maritima* whose long delicate tendrils shelter quantities of small crustacea. Since the latter are also the main food of the coots and wintering ducks, the flamingoes are in direct competition with these and drive them away. Noisy birds when feeding, the flamingoes whoop like swans, honk, quack and grunt incessantly. From time to time there is a general raising and lowering of those long Roman-nosed question-marks which represent heads and necks, and perhaps a spreading and flapping of those breathtakingly beautiful scarlet and black wings; while two may stretch up their sinuous necks, beak to beak so that the recurved hook to the upper mandible lends them a curiously supercilious and parrot-like profile. At frequent intervals individuals make short flights, with stepping toes pattering over the glittering gray wavelets, from one feeding group to another. They are exasperating birds to watch in the winter, being ever on the move; and with such vast areas of shoals in which to feed, they soon pass out of observational range as they follow the slowly ebbing waters.

But consider these same flamingoes in the summer. All through the dry season the *étangs* and lagoons are exposed to near tropical heat, though temperatures rarely reach those in the *marismas;* a heat rendered more intolerable by the mistral which sweeps away the last remaining drops of saline water. "Words fail to describe this vision of endless salt when evaporation has replaced the water by a vast sheet of glittering crystal" wrote Étienne Gallet, who solved the problem of the flamingoes' erratic breeding in the Camargue. "The minute fauna of the brackish lagoons is dead and dissolved in water. Only a few rare living organisms, hardly visible under the microscope, manage to live on in the deadly brine . . . even the brine-shrimp perishes." Nevertheless, from this Dantean salt desert the thousands of flamingoes breeding in the Camargue obtain fully adequate nourishment, as they also do in the *marismas.* When it wants to eat, the flamingo walks backward with its head under the water, say the marshmen of the Camargue. In fact, within the restricted space permitted by the thousands of its fellows from the nearby nesting city, a flamingo paddles around in the mud in one place, pivoting on its feet, while at the same time describing a circle with its long neck and beak. This compass movement with the feet works the mud and sand into a circular mound, upon which the flamingo deposits the rejected coarse sand, from which its lamellae have filtered the food contents, forming what Gallet terms appropriately a "cone alimentaire." These cones are a characteristic feature of the brine lagoons adjacent to a city, and Gallet describes how the flamingoes, ceaselessly masticating their revolting mud, continue to trample out these small feeding hollows throughout the summer. Only the constant churning of their feet keeps the fluid black mud liquid and prevents it from crystallizing into salt, and should they cease feeding in a particular place for a day or two, a crust forms almost immediately.

But what food can the flamingoes obtain from this unpromising material? From the evidence of the few stomachs that have been examined, it would appear that, from the coarse sand, the Camargue birds are actually filtering mud with an organic content of 6 to 8 per cent. Moreover, an American research biologist, Paul Zahl, observed that both on Andros and Inagua some herds of immature flamingoes were feeding exclusively in large brackish lakes in which there were no cerithiums, and no other source of food except the extraordinary bacteria *Halophilus* which can only exisit in a solution of concentrated brine, and whose pigmented myriads sporadically stain the evaporation brine-pans pink or red. It would seem therefore, that flamingoes are capable of subsisting on pure mud, though their method of filtering this must obviously be different from that used when feeding on brine-shrimps or cerithiums.

Charles McCann, who visited the Great Rann of Cutch in the late 1930s, concluded that since there are no invertebrates, nor aquatic vegetation on the Rann, the flamingoes, both adults and young, must feed on the seeds of such sedges as *Scirpus maritima* and *Rupia rostrellata*. Neither of these grow in salt water, but quantities are flooded on to the Rann by the rivers from the Kawra marshes some 65 miles distant, and are able to seed and die on the Rann before the seasonal inflow of the sea. But it is most unlikely that sufficient seeds could be washed on to the Rann in this way to support half a million adult and young flamingoes, and it would seem more likely that they too feed on organic mud during the breeding season. Another calculation by Leslie Brown, on the assumption that these flamingoes would require 10½ ounces of food per head per day, indicates that these half-million birds would consume 21,750 tons of mud during the five months of the breeding season.

We have seen that whether flamingoes establish a nesting city in a particular year in a particular locality, provided that the food supplies are adequate, depends primarily on the level and salinity of the water in the lagoons; but successful breeding also depends on a third factor, most graphically expressed by Abel Chapman's often misquoted saying of the *marismas'* herdsmen: if the shadow of a cloud passes over them, they forsake. During the period of nest building and incubation flamingoes are extraordinarily nervous, though this nervousness abates as the hour of hatching approaches, increasing again after the chicks have hatched. Gun-shots, dogs roaming in the immediate vicinity of a city, indiscreet visitors and photographers, and particularly low-flying aircraft are all sufficient to cause the flamingoes to desert their city *en masse*.

One other factor has to be taken into account in the establishment of a nesting city: the mud must be suitable for building material, soft enough to be used as mortar. Indeed, in the Camargue, where the flamingoes build their cities exclusively in the salt-pan area, in which there is little variation in the water-level because the influx of sea-water to the evaporation pans is controlled by sluicegates, it is the condition of the mud and not the level of the water that is the decisive factor in the choice of a particular site. In Gallet's long experience the flamingoes breed in these brine-pans even in drought seasons, because the thick crust of salt protects the organic mud from the sun's rays and keeps it moist: whereas when the natural saline lagoons have dried out, their mud bottoms, washed by fresh drainage water, lack this protective covering and harden and crack in the sun. Only when no bare clayey ground, which water permeates but hardly covers, is available, do the Camargue flamingoes build their cities on submerged sites, though in these circumstances on higher islets, despite the fact that these may carry too luxuriant a growth of saltwort. However, once a suitable site has been determined on, building operations are conducted in a frenzy of activity by the whole colony of several thousand birds, and Gallet has described how in a single day, or a single night of exhaustive labour, a colony of flamingoes is capable of converting an islet covered with saltwort into a city of mud cones. Trampling and tearing at the tufts, nipping them off, digging them out at the roots, they totally denude the islet of vegetation and overlay it with the rudimentary outlines of nests. The fact that this ruthless efficiency eventually results in the destruction of the islet as a permanent breeding site may also be a factor in the flamingoes' frequent removal from one locality to another, for the saltwort scrub protects low-lying islets from the winter rains and from gale-lashed waves; but when these are denuded by the flamingoes they are quickly flooded, hollowed out and carved up into secondary islets, whose elevation is too low for saltwort to recolonize and to survive the summer when the water is saturated with salt.

The flamingo's building technique is an elaboration of the feeding technique, for while scooping away the viscous black mud with beak and feet, the bird rotates and in so doing excavates a circular trench with a circumference of 8 or 9 inches, piling up the mud, together with any vegetation present, on the ever-growing central mound. (Its incessant trampling incidentally moistens the soil additionally by assisting the water to flow, by capillary attraction, to other inaccessible places.) The end-result is a truncated cone, resembling an inverted flower-pot or a potter's-wheel cone, whose height depends upon the consistency and adhesive properties of the mud used. Some cones are hardly more than slight elevations of the mud, but on suitably moist clay the mound is about 8 inches high and equal to the depth of the surrounding trench, the outer rim of which overlaps that of neighbours, so closely do the flamingoes build in groups or *vetones*. However, in some parts of their range a series of favourable seasons may allow the flamingoes to return to the same city year after year, and both McCann and Salim Ali, a subsequent visitor, reported that on the Rann of Cutch the same nest cones had been repeatedly used, and varied in height from 2 inches to 2 feet according to their usage and the depth of the water.

It is possible that in those parts of the world where water-levels are more variable than in the Camargue, the flamingoes adjust the height of their nests to comply with seasonal conditions; and F. M. Chapman, in describing the thousands of nests built in the mangrove swamps on Andros Island in 1902, stated that since the flamingoes were exposed in this habitat to the danger both of floods after tropical downpours and of tidal flooding, they built their nests to an average height of 10 inches, which was sufficient to keep the eggs above the water-level. But in this he may have been mistaken, for fifty years later Paul Zahl noted that when the nests ot these Andros flamingoes were flooded by high tide, they removed to higher ground and built a second series of nests on the solid coral rock. Nor is there any confirmation from other localities of flamingoes adopting preventive measures against flooding. The Kara Bogaz Gulf city on the Turkmenian shores of the Caspian is built on moist saline silt covered by a hard crust of salt mixed with cadmium, which the birds are reported to crush with their beaks and in some instances ram with their feet. However, the bases of their nest cones lie in shallow water and, if the wind blows off the sea, are flooded. Russian sources do not state what action the Kara Bogaz flamingoes take in these circumstances, but in other localities flooding results in the colony deserting. Prior to 1962, for example, Lake Natron was the headquarters of the 3 million lesser flamingoes in East Africa, but as a result of persistently high water-levels, almost filling the lake that year, they migrated the 30 miles to Lake Magadi. There, between June and October, more than a million pairs, together with 10,000 greater flamingoes, built their cities on the soda deposits. During March and April the following year, however, some 2 million of the lesser assembled on Lake Nakuru, 80 miles to the north, where they could be seen packed solid in a 50-yard band along 2 or 3 miles of the shore-line. Again, a continued rise in the waters of this lake, to a level as high as in living memory, resulted in all leaving by July, and it was not until October that 40,000 nests were located in a freshwater lagoon at the southern end of Lake Magadi. However, when exceptionally heavy rains flooded this colony too in November, almost all the inhabitants of the city deserted, abandoning both nests and eggs.

An additional cause of desertion by these most temperamental of birds arises out of their compelling urge to incubate in close-knit groups, and it has been conjectured that as high a proportion as 20 per cent may abandon late clutches of eggs because earlier hatchings have resulted in gaps of a few yards being opened between them and their nearest neighbours. This habit makes it all the more remarkable that

Flamingo with chick on nest cone

flamingoes can breed successfully in very small colonies on such oceanic islands as Aldabra and five of the Galápagos; but as long ago as 1684 Ambrose Cowley reported in *Voyage Around the World* seeing a couple of dozen flamingoes on a shallow lagoon on the Galápagos island of James, while in 1949 Wolfgang von Hagen photographed chicks in process of hatching from a colony of less than twenty nests in a small saline pool surrounded by giant cacti on the small island of Jervis.

Flamingo chicks hatch after a month's incubation, and for the first fortnight are fed by regurgitation with a clear white liquid resembling water. Its composition has not, I believe, been established, but it may possibly be saliva. The parents continue to feed the diminutive chicks with this liquid, though increasingly sparingly, until they are eight weeks old. Nevertheless, even newly hatched chicks devour with astonishing avidity anything remotely edible, including withered marine grasses, shell fragments and, incredibly, in the Camargue, large quantities of highly saline baked mud in which the most primitive invertebrates cannot exist, and when a couple of weeks old they are feeding like adults, despite the fact that at this age their beaks are of conventional shape and have barely begun to curve.

Consider the conditions in which these young flamingoes will have to survive on the African soda lakes and in the salt deserts of the Camargue and the Great Rann until they fledge when ten weeks old. At midday in the scorching heat and blinding glare of Lake Natron's inferno the temperature of the city's sodium surface soars to 150 degrees F or higher, though for the first few days the chicks are able to evade the most extreme heat by remaining on top of their nest cones where the temperature is only at blood-heat, and to these they always return if they have been frightened off them. Nevertheless, that conditions on the soda lakes can be dangerous for young flamingoes, once they leave their cones and walk through the soda-saturated water, is indicated by the fate of those that hatched in the Magadi city in 1962, for large numbers of these became so encrusted with hard ball-like masses of soda around their ankles that they were dragged down into the mud by the weight of these shackles and immobilized. Many died, but many thousands were rescued by being driven into less saturated waters and their fetters knocked off. So too, when in June 1969 the Etosha pan dried out and more than 30,000 young were deserted by their parents, these had to be transported to a flooded pan 30 miles distant—a rescue operation which, though achieved with a loss of less than 3 per cent, was, as we shall see, possibly unnecessary.

Conditions in the Camargue are not normally quite as inimical as those on the African lakes, but in drought seasons, when the almost total evaporation of the water in the brine-pans and lagoons has left only crystals of salt covering the thick black mud, the young flamingoes are obliged to exist in a hyper-saline environment. In his beautiful book, *Camargue: The Soul of a Wilderness*, Karl Weber has described what conditions are like in the Camargue when no rain falls for three or four months:

> Day after day the sun beats down from a pale cloudless sky with undiminished strength—scorching, crushing, killing. . . . Mirages, shimmering reflections in the air over the parched ground, suggest the proximity of non-existent water. The small and medium-sized ponds have long dried out; ditches and water-holes have been empty for weeks, their banks covered with a yellow-brown fringe of wilted bulrushes and sedges.
>
> Even the large inland lakes are rapidly losing their water . . . the exposed belt of mud dries out rapidly, hardening into a . . . grey mass. Wherever the water lays bare mud and earth a bizarre lattice-work of fine cracks and fissures opens on the

crusted surface. . . . The large brackish lake which we had crossed in mid-May . . . whose water then came very nearly up to our hips, has disappeared. In its place stretches a grey desert plain, extending far into the haze of the distant horizon, its silent infinity varied only by occasional shimmering white efflorescences of salt or the bleached armour of dried-out crabs.

The low salicornia plants along the paths and tracks are covered with a thick layer of yellowish-grey dust. . . . As soon as the wind springs up huge clouds of fine soil are driven across the plain, coating shrubs and plants and even dusting the tall tamarisks with a white layer of powder. Hundreds of *Cepaea* snails have climbed up grasses and shrubs, bushes and fencing poles. . . . On the parched ground they would now be slowly scorched. They cling together in huge clusters like grapes, or like twisted chains.

The days pass in dull heat, heavy and monotonous. The air trembles over the hot ground, the hard light steeps the landscape in a dull grey-green haze. Now is the worst time for water creatures, especially the fish in marshes and ponds. The steadily dropping water-level increasingly compels them to congregate at the few remaining deep spots. . . . The salt content of the water increases, soon the oxygen will no longer be sufficient. . . . The first fish leaps out on land to die there in the sun. A feast has been laid out for herons and kites. The herring gulls, too, those hyenas among the birds of the Camargue, are not long in coming. Circling at great height over patches of water, gliding slowly before dropping down to begin their meal.

It is under these appalling conditions that the chicks, when a fortnight old, form themselves into homogenous age-groups for ten days or so. This is a particularly critical period for them, because any disturbance of the city results in groups of different ages mingling, with the result that they are not able to present a solid defensive phalanx against their few predators in the Camargue. These, somewhat unexpectedly, are predominantly herring gulls which, breeding in small sporadic colonies, with one or two pairs nesting on the fringes and even in the middle of the city, take both chicks and eggs, just as scavenging vultures take a regular though not very large toll of eggs from the Great Rann city, and fish-eagles prey on the young flamingoes of the African lakes. But once this critical period is past, all age-groups unite in a single colossal herd, several thousand strong, which is accompanied (as in the case of emperor and king penguin chicks) by a few adults. The young can now run fast and swim well, and are safe from all predators, though the herring gulls continue to prey on weaker stragglers, unable to keep within the ranks of their fellows when winds whip up waves on the shallow lagoons; for as the saline waters evaporate, so the flamingoes retreat to the waters of the Étang de Vaccares which, constantly replenished by fresh water, never dries out. So too, as the waters recede from the Great Rann, the city's immense herd of chicks, accompanied by only two or three or half a dozen adults, marches in their wake, and may continue to do so for weeks until able to fly. Their line of march is signposted by the corpses of hundreds of weaklings that have succumbed to the strong desiccating winds associated with noon temperatures exceeding 116 degrees F.

Where their area of movement is restricted these enormous herds of young can themselves be responsible for immense damage in their city. At the Lake Natron city maribou storks are constant attendants. Although these scavengers never attack the adult flamingoes, and only rarely the chicks, Brown observed that their forays resulted in inexplicable panics among the incubating flamingoes, leading to temporary desertion on a colossal scale. Moreover, once nests had been deserted, further destruction of eggs and newly hatched chicks resulted from the extraordinary behaviour of the young of earlier hatchings who, driven by some uncontrollable urge, milled over the city's island site like an army of automatons, trampling everything that lay in their way. These aimless and anarchic maneuverings are presumably the early manifestations of the instinct that will eventually lead them to march, often for miles, to the nearest open water in the middle of the lake, for they do not migrate to the shores until almost ready to fly. The fact that so few adults accompany these juvenile herds does not nessarily imply that the parents have deserted their young. When, for example, the Magadi city was abandoned during the November rains, not only nests and eggs were deserted, but also 4,000 or 5,000 chicks. But after the city had again been totally immersed on the last day of the month these chicks were found to be alive and thriving, for their parents were returning in the evening, from the direction of Lake Natron, to feed them. When the adults left before daybreak the last 200 to leave were seen to fly over the chicks and herd them into a tight swarm, before also departing.

Despite the catastrophes resulting from the peculiar nature of their environment, their irregular breeding and their frequent mass desertions, flamingoes being long-lived birds, with a life expectation of certainly more than twenty years, raise sufficient numbers of young to maintain their various populations. Although, as we have seen, the Camargue flamingoes established only sixteen cities, containing from 4,000 to 8,000 nests, between 1914 and 1947, four of which were destroyed or partially destroyed during the incubatory stage, and one during the chick stage, they nevertheless reared some 39,000 young during this thirty-four-year period. Similarly, the East African herds of lesser flamingoes based on Lakes Natron and Magadi, though also not nesting annually, have reared an average of 130,000 young per nesting year over a period of ten years.

During the hot months of the breeding season flamingoes seem able to tolerate all extremes of temperature. In the salt and borax deserts of the High Andes, for instance, which are frequently covered with a little water in the summer, the temperature can span 100 degrees F in twenty-four hours, rising from −10 degrees during the night to +90 degrees among the rocks at midday. But throughout the cold winter no rain falls, and the flamingoes emigrate as the *salinas* dry up. Similarly, two-thirds of the Camargue breeding population emigrate before the winter, presumably to Africa where ringed birds have been recovered in Mauretania. When the Camargue experiences its occasional severe winter the resident one-third suffer very high losses; but again that mysterious built-in survival mechanism (apparently common to all forms of life) operates, enabling the survivors, together with those that have wintered in Africa, to breed exceptionally successfully the next summer. After the unusually hard winter of 1955–6, for example, when 1,800 flamingoes died, the largest number of young ever recorded in the Camargue were reared the following summer; while after the seven-year hiatus from 1962–8 no fewer than 7,300 nests were built in 1969 and some 6,000 chicks fledged successfully.

According to Brown, the flamingoes of the African lakes are preyed upon not only by several kinds of eagles and vultures but also by a variety of carnivores from ratels to lions, while large numbers of chicks are killed by such scavengers as hyenas. He suggests that this vulnerability to predators explains why they build their cities in the terrible conditions of heat and glare in the middle of the lakes, where they are safe from all but a few of the raptors, except in those years when the mud dries out sufficiently for carnivores to reach them from the shore. So too, when the water-level of Inagua's salt lake drops, the crop of nestlings is wiped out by herds of feral pigs raiding across the hard dry mud. But in other parts of their range flamingoes apparently have few enemies and, being admirably equipped to survive the peculiar hazards of their unique environment, are threatened only by man's activities. Their cities in the Andes are, for example, regularly plundered for eggs, and the birds themselves trapped and hunted with bolas, despite their remoteness which resulted in a race of lesser flamingoes, first discovered in 1886 on the saline lakes and lagoons in the snow-covered mountains beyond Chile's Atacama desert, not being rediscovered until 1956. In that year a city of several thousands was located at an altitude of over 14,000 feet in a habitat remarkable even by flamingo standards, for wind-rows of ice, feathers and salt crystals mingled on the lagoon's beach of black sand, while across the middle of the lagoon high white salt-cliffs separated water stained brick-red with algae from the inflowing blue streams of subterranean hot springs and melting snows. In the Caribbean the flamingoes are harassed by low-flying air-craft, while such large numbers have been killed by the local inhabitants that their total population has been reduced to less than 20,000 from five times that figure. In the Camargue the extension of rice cultivation presents a threat, though this has been offset by the World Wildlife Fund's recent acquisition of a part of the salt company's land. And if it be difficult to envisage any industrial or military encroachment on their intolerable habitats on the quicksands of the Great

Rann of Cutch or on the salt deserts of the Kalahari or the soda-lakes of the Rift Valley, there is in fact a soda industry at Lake Magadi—though this at present extracts less than the inflow of salts deposited by the alkaline springs—and there is already an invasion by tourists of another of the lakes. It is in the nature of a miracle that a bird with such specialized requirements should have survived so long.

15: Wading Birds on Northern Mudflats

As on tropical shores, so on those of temperate seas the daily lives of hosts of creatures from burrowing worms to wading birds, duck and geese are controlled by the rhythm of the tides, whether their habitat be the sandy beaches and mudflats of the coast or the mudbanks, sandbanks and creek-cut salt marshes many miles inland up the estuaries. At low water every square yard of exposed mudflat is imprinted with the tri-pronged slots of wading birds, from diminutive stints and dunlin to huge curlew, and with the webbed triangles of duck, especially wigeon, mallard, teal, and shelduck. All are attracted by the teeming life of the between-tides zone. They represent the upper limit of the food-chain which begins with the plankton, for they have no significant predators except man the wildfowler, though the few remaining peregrine falcons or duck hawks, now almost exterminated in

North America, hunt over the coastal mudflats and estuarine salt marshes during the winter months. When, in Britain, the sea flows in over a Northumbrian slake, flooding up the furthermost creeks at high water, flock after flock of curlew and godwit and smaller waders rise from the last mudbanks to be covered, and flight over to a rig of sand and rocks which always remains dry except at the highest spring-tides. Ultimately several thousand waders are massed on this last dry refuge above the tidal flood. The peregrine falcon is aware that the waders retreat to this rig at the hour of high water, and that many duck—wigeon, merganser, eider—frequent the shallow waters at its edge. Day after day two peregrines hunt together over the rig, and the flock of 1,000 or 3,000 fearful godwit, flushed into flight, perform marvellously intricate escape maneuvers, "whiffling" down time and again from the dreaded stoop of one or other of the peregrines, before ultimately streaming out over the slakes in bunches and chevrons, sweeping and twirling down to the water, stringing out just above the waves, and then soaring up again.

But when spring-tides are whipped up by gales, then every acre, every square foot of slake and rig is covered at high water, and the waders' only retreat is a long narrow strip of sand, a yard or two wide, at the base of the sandhills. To this strand large and small flocks of waders fly in from all quarters. But no sooner have they assembled than they are impelled, by a common and irresistible excitement at the presence of so many of their kind, to give free rein to that supreme expression of all their moods—flight; and 5,000 curlew and godwit, rising as one, go rushing out over the breakers. Soaring, stooping, changing direction with a hardly credible instantaneousness, they split into separate flights and come together again, while weaving further and further out to sea, until lost to sight in the blown spray's mist that ever rises from the breakers on this storm-bound coast.

But 10,000 dunlins, together with some hundreds of knots, gray plover, and turnstones, remain on the strand, and when the combers are surging far up the face of the sandhills further along the shore, tearing down great slabs of cemented sand and exposing the fibrous network of the long marram-grass roots binding the dunes (in which rock pipits and linnets will fashion their nests in the spring), these smaller waders swarm in insectivorous masses on this one dry strip, feeding avidly on the millions of flesh-fly maggots that have been churned up by the huge seas from the broad banks of rotting seaweed and tangle, unwetted since the last spring-tide. Dibbling in the sand at the breaking edge of the tremendous sea, the dense drab carpet of dunlin seethes with ceaseless activity, as all scurry hurriedly up the slope of the sand shelf from the sudden curl of a breaker, and surge down again in the very waters of its hissing backwash. Some perhaps lag behind their fellows for a vigorous bathe in the foaming flood of a spent wave, and then hop up, one-legged and without even bothering to unbury their heads from their scapulars, when a second wave threatens to submerge them. And then, when the sea begins to go back, the thousands are once again delving for dear life, inches apart, in the moist shining sand for their microscopic prey.

The hour of high water has passed. The sea ebbs from the sands, and one mudbank after another is uncovered. The armies of waders split up into smaller feeding flocks and fly off, each to its special flat or pool or creek. Most blessed among birds, their food is guaranteed in abundance except during periods of low tides in the most severe winters when even the mudflats are iced-up. The density of life in this inter-tidal zone, though composed mainly of minute insect larvae and young bivalve mollusks, periwinkles, and crustacea, is greater than in the richest farm soil, amounting to as much as a quarter of a million individuals to the square yard. Such are the numbers of lugworms that, when the tide goes back, every 100 acres of mudflats may be spattered with 4½ million worm-casts, the residue of 900 tons of mud and sand they have displaced; while a single square yard of sandy shore between tides may contain from 1,000 to 8,000 tellins: those ½-inch-long bivalves with flattened oval shells, smooth, delicate, translucent and tinted pink, orange or yellow. The tellins are a favourite food of waders with medium-length beaks, such as redshanks.

On sandy beaches, too, the numbers of small amphipods, collectively known as sandhoppers, are so great that they have been reckoned in cartloads rather than in millions of millions. Burrowing, in the morning, under the tide-rows of seaweed piled high up on the beach, the hoppers emerge in the evening to "jump for joy," propelling themselves several feet by a sudden straightening of their normally bent shrimp-like bodies, possibly in pursuit of their minute planktonic prey. But though reputed to be nocturnal, any disturbance of the seaweed debris sets them jumping *en masse* at any hour of the day; and though also reputed to drown if submerged for long periods, they, and their relatives the shore-skippers which live under stones near the low-water mark, are common enough on the wet sand at the edge of the sea. There they are snapped up by every kind of wader, but especially by the small trips of sanderlings, pattering along the very brink of the waves, and indeed often into the foam of receding ones, in order to retrieve hoppers or shrimps before they are sucked back by the undertow.

On clean sands, where they live in maximum densities, 100 acres may hold up to 150 million edible-cockles in burrows just deep enough to cover their shells and the two short siphons which project above the water when the tide is up. Cockles, however, like many other inhabitants of the tidal zone, are liable to decimation in very hard winters and also by sudden rises to high temperatures in the late spring. Moreover, whole beds of cockles have the disconcerting habit of occasionally migrating to new beds. It is true that a cockle's large, rounded and fleshy foot, bent in the middle, provides it with a limited power of locomotion, for protruding its foot above the surface and pressing its tip against the sand, the cockle suddenly straightens it, as if releasing a spring. This impulsion is sufficient to cause the cockle to roll over several times or even skip into the air if the jerk is strong enough. But though the image of a colony of cockles skipping over the sand on migration is irresistible, one supposes that such migrations are in fact set in motion by currents.

At low water one can also tread delicately over carpets of millions of young edible-mussels covering acres of mudbank or reef. Indeed no shore animal grows in such dense masses or can establish itself as quickly. C. M. Yonge has described how in 1944 the Dutch, in the course of their campaign to free their country from the Germans, breached the dykes protecting the island of Walcheren, with the result that the greater part of the island was flooded with sea-water. When the breaches were closed and the water pumped out about a year later, road surfaces and the sides of houses and fences were found to be covered with mussels, which even hung in clusters like fruit from the branches of trees.

Every species of wader, and indeed individuals within the species, has its particular food and method of obtaining it.

Dunlin, like sanderlings, will dash into a receding wave to retrieve sandhoppers but, as we have seen, more commonly swarm like mice over the sand or mud, incessantly tapping the surface or probing up and down in it with their short black bills, burying them up to the hilt. When feeding on minute mollusks, such as the ⅓-inch-long laver spire-shell snails buried only just beneath higher levels of the mud, the dunlin appear to take only those that make a slight movement; but when amphipods and worms are numerous, secure these by dashing forward and plunging in their bills, possibly when detecting their vibrations. They do not pull up a worm here and a worm there in the inconsequent manner of ringed plovers, which usually feed singly and never in large flocks. With that technique characteristic of all plovers, a short quick run and a dip forward, the ringed plover feed higher up the beach than other small waders, and often above the level of normal high tides. Indeed, they never seem to settle to a sustained bout of feeding on the shore, though at high water of spring-tides and in stormy weather as many as seventy often accompanied by a dozen of the larger gray plover, may flock together daily to the smooth fairways of a sandy links, a few hundred yards inland. Honeycombed with rabbit warrens and also covered with ragwort plants, the links are the habitat of small helix snails, which the plover pick up with unusual industry.

There is an explanation for this apparent watchful hesitancy of feeding plovers, whether ringed or gray, golden or green. After pulling up a sand-worm, a ringed plover stands motionless, apparently listening, then turns completely round and darts away to extract another worm several yards distant. Observe, too, a gray plover feeding. It stands tense and motionless—and, if it is not listening for worm tremors in the mud, its attitude belies it—then suddenly darts forward, nearly always forward, though its angle of vision is lateral. At the end of its run, often of some yards, it usually extracts a worm without delay, though it occasionally stands bowed forward over the spot it has selected, before finally uprooting the worm. There can be no doubting the efficiency of its technique, for one will pull up the long thin lugworms at an astonishing rate, extracting a fresh one every few seconds. The majority are pulled up smoothly and easily, though a stubborn worm may require a good tug and a second or two's straining back on the part of the plover. These short-billed waders then, unlike the longer-beaked ones, would appear to feed mainly by ear or by detecting vibrations, whereas godwit and snipe, probing vigorously in the ooze, appear to feel for worms with the sensitive nerve-ends to their extra-long beaks, though perhaps they too are picking up vibrations.

Lugworms (or lobworms) and ragworms are the main food of such very long-billed waders as godwit, whose slightly upturned beaks are more than 3 inches long, and curlew, whose impressive sickle-shaped beaks measure from 4 to 6 inches, though they look much longer than that. Ragworms, which are bristle-worms from 2 to 18 inches long, inhabit burrows under stones and boulders, and with the aid of the bristles protruding from their body segments can crawl actively and, in some instances, swim. Large ones can bite painfully, though not, unfortunately, sufficiently painfully to deter anglers from digging them for bait in such quantities that they have been virtually exterminated on many stretches of British coasts. Their extermination must in due course affect the ecological structure of the mudflats.

Lugworms are 8 or 9 inches long and as thick as pencils. The commonest of all the sand-worms, their habitat is mainly sand, but since they feed on the small amounts of organic matter contained in the soil, they are also to be found in areas of sand mixed with mud. It is the lugworms that are responsible for the multitudes of worm-casts thrown up after every receding tide. Each cast rises beside a circular depression in the sand, for the lugworm lives at the bottom of a double-shafted U-shaped burrow, the walls of which it strengthens with mucus, so that it can be said to occupy a tube within the sand. The tube has two surface openings. Through the head-shaft (at the circular depression) the worm draws down mud and sand from which to extract food, while down the tail-shaft (at the worm-cast) comes a continuous current of water for respiration, driven by waves of contraction along the worm's body. The danger of oxygen deficiency when the sands are dry and the burrow only partly filled with stagnant water, is probably avoided by the storage of oxygen in the worm's blood, and also by a bubble of air trapped by its tail-end and brought into contact with its gills. From time to time the worm backs up the tail-shaft in order to defecate digested mud and sand in the form of casts on the surface, and even the extremely long-billed curlew and godwit can only reach the lugworms when they ascend into the vertical sections of their burrows for this purpose.

Godwit, though also feeding on small crabs and mollusks, shrimps, and fish fry, are especially partial to lugworms and are to be found daily feeding on sandy stretches rarely visited by curlew. The sands of a small cove may indeed be laced from end to end with their trident footprints, and pitted here and there with awl-shaped holes, ¾ inch across, where they have pivoted round 360 degrees on boring bills. Those feeding in shallow pools in the lagoons shove their heads right under water, stabbing the oozy bottom violently, and frequently raising their heads to swallow morsels or to shake their pinkish bills while ejecting a stream of mud. It is surprising that their bills do not become choked with mud.

The wintering population of godwit do not feed in one composite body, but split up into small flocks which frequent various parts of shore and slake, coming together at high water. In autumn, winter and spring, and from one year to another, a flock of a dozen or of two or three score is to be found day after day in certain small sandy bays, where oozy

Black-tailed godwit

lagoons, rock-pools, reefs, sandy shallows, rotting storm-cast banks, and ramparts of brown, pink, orange and green tangle and seaweed provide a favourite feeding ground for various species of waders and also for rock pipits and starlings, and for snow buntings and larks when the hinterland is frozen. Turnstones also feed in these banks, and their technique is different from that of any other wader, for prying energetically among the tangle, they lunge headfirst into a clump of it, shoving it up and away from them with their blunt heads and short conical upturned beaks, exposing the sandhoppers and small mollusks hiding beneath. So vigorous is this action that they often push their heads right through a clump, as they shovel it to either side with a little run. Sometimes a number of turnstones may combine to turn over a large dead fish, first excavating the sand on either side of it in order to obtain a purchase, while migrants to such tropical wintering places as Hawaii have been reported eating the eggs of colonies of sooty and gray-backed terns. This is such an unexpected departure from the turnstones' normal routine that it is worth quoting from Alexander Wetmore's account in A. C. Bent's *Life Histories of North American Shore Birds*, in which he describes how:

> As we moved through the great colonies of sooty terns, the birds near at hand rose before us from their eggs, often communicating their alarm to neighbours, so that at times clouds of birds arose to fill the air. At our heels, fifteen or twenty feet behind us, came little groups of turnstones. . . . The turnstones ran quickly about, driving their bills into the eggs without the slightest hesitation, breaking open the side widely and feeding eagerly on the contents, sometimes two or three gathering for an instant to demolish one egg and then, with one half-consumed, running on to attack another. . . . The densely

packed colonies of aggressive sooty terns were open to attack mainly along the borders . . . but little scattered groups of grey-backed terns on the open beaches were entirely at the mercy of the turnstones. . . . So bold were the turnstones that on one occasion I saw two actually push aside the feathers on the sides of the incubating tern, drag her egg from beneath her breast, and proceed to open and devour it within six inches of the nest.

Turnstones normally associate in smallish flocks of two or three score, but at mid-August when they are returning from their breeding grounds in the far north, an immense assemblage of as many as 2,000, accompanied by 150 purple sandpipers, may gather on the reefs of such sea-girt islands as the Farnes, which stretch from 1½ miles to 4¾ miles off the coast of Northumberland. Reefs are a typical turnstone habitat, for though they can be found feeding on almost any stretch of sand or mudflat at one time or another, sea-weed-covered reefs, rock pools heavily tangled with the broad oar-weed, and mussel scaps are their special feeding places. And these are the invariable habitats of their constant companions, the purple sandpipers which, while picking minute winkles and crabs out of crevices in the rocks, or gobies or fish fry from the pools, venture right under breakers curling over the reefs. With that lack of fear, inexplicably displayed by some species of birds but not by others, purple sandpipers feed confidently within a few feet of man.

A greenshank displays a similar explosive energy to that of a turnstone, when feeding in a creek well up on the mud-flats, though its technique is quite different. Dashing through a shallow pool, with neck extended and the fore part of its head immersed, a greenshank pushes its long upturned bill before it with a smooth though slightly impeded motion, as if forcing it through the ooze and water, while in pursuit of a shrimp or prawn or small crab; or spins around suddenly in sweeping circles after a small fish that has been isolated in the pool by the ebbing tide. A redshank, in contrast, immersing its delicate sharp bill and three parts of its head in the same shallow pool, dibbles from side to side with extraordinary rapidity, often gyrating its body around its bill in a complete circle. Then, wading belly deep or swimming across the pool, it probes in a hundred different places in the ooze for laver snails or small cockles.

That gigantic long-legged wader, the gray heron, is also a member of the mudflat fauna, though also of course of inland rivers, lakes, and ponds; but though it may gobble up a crab now and again, the fish and eels in the pools and creeks are its primary concern. The popular conception of a heron poised motionless, breast deep in a creek, spearing fish at hourly intervals, is false. Rather does it work fast, stalking purposively, if cautiously, with S-shaped neck coiled to strike, from one fishing place to another. A grave slow-moving bird perhaps, until that keen yellow eye detects a fish, when the sinuous neck uncoils and the wings lift a little excitedly: a lightning thrust of stabbing beak, and a slapping fish is writhing between the sharp mandibles, which mash up the small bones; tilting back its head, the heron swallows the fish with a sudden gulp. But if it has captured a large fish, it stalks with characteristic deliberation from the creek and beats the fish two or three times on the bank before swallowing it, or perhaps places it on the bank and picks the flesh off the bones; the heron then cleans its beak thoroughly on a clump of weed, and paces down again to the creek to drink. Eels, the heron's speciality, are beaten almost to a pulp before being swallowed, not only being pecked and pounded on the ground but, in the words of that fine field naturalist of the old school, Lord William Percy, subjected to a prolonged course of treatment that can best be likened to the action of a man repeatedly cracking a hunting crop. Each time the eel is cracked its loose swinging end is whipped against the heron's head and neck, and by the time the treatment is concluded the heron's feathers have become fouled with slime. It has been said that much of a heron's time, particularly when on nest duty, is passed in cleaning and waterproofing its feathers with the aid of powder-down based in thick greasy yellow pads in the skin—one pair of pads on the upper part of the breast, a second pair on the back of the thighs, and a third pair on the lower side of the abdomen. In Percy's experience, powdering, combing, and preening appeared to be the most important operation in a heron's life:

> The location of the "powder-puffs" is such that when the bird is making use of them, as the head passes over the upper (breast) puff down to the lower (thigh) puff, the neck is brought into contact with the upper while the head is being rubbed on the lower. . . . Each time the head emerges it is seen to be increasingly covered with powder and dotted with particles of the bluish powder-down itself. The degree of powdering depends on the degree of the contamination . . . and it will therefore range from a perfunctory rub-over to a thorough powdering, combing and oiling, occupying a couple of hours or more. In the latter case the process consists of three well-defined stages:
>
> 1, powdering; 2, removal of the powder together with the eel slime by the use of the comb on the middle toenail; 3, application from the oil gland of the waterproofing of the plumage which has been removed together with the slime. If there is much damage to be repaired, the first stage may be

interrupted by a pause to allow the powder to dry off and then be repeated all over again.

Curlew, though feeding on worms, velvet-crabs, shrimps and such fish as small flounders and blennies, are in particular shellfish merchants. When the tide has ebbed back, pack after pack will glide into the mudflats and corkscrew sharply down, until some 3,000 are ranked up on the mud, turning over stones and weed in search of winkles, bathing in shallow dubs, or idly titillating sodden leaves with the tips of those immense bills, then letting them fall and allowing the wind to blow them away again. Mussels, cockles, winkles, and small clams are all acceptable to curlew, though they are especially partial to peppery furrow-shells, which are bi-valves with very flattened oval shells between 1½ and 2 inches long. Burrowing 6 or 8 inches deep in fine organic mud in higher levels of the inter-tidal zone, where the salinity is lowered by fresh water, the furrow-shells' presence is revealed, when the tide is out, by the star-shaped imprints on the mud of their two mobile siphons, up to 6 inches long, which extend in all directions when the tide is up, while ceaselessly sucking in food.

Though oystercatchers, like other large waders, probe for worms, with a very rapid up and down action resembling that of a sewing machine, and locate them by touch, they are the shellfish merchants in chief; and their winter habitat is centred around mussel-scaps, cockle-bats, reefs, and flag-stone flats studded with conical shells of limpets and densely populated by winkles under every boulder and clump of weed. There is an infinite variety of food for the taking, and the oystercatchers have evolved efficient techniques for dealing with their various prey, though whether one oystercatcher is equally efficient at dealing with worms, crabs, cockles, mussels, and limpets, or whether every oystercatcher tends to specialize in particular types of prey, would prove an interesting field of study. Having, for example, flipped a crab over on to its back and destroyed its central nervous system with a stab in the mouth, an oystercatcher then hammers its orange-red beak into the crab's abdomen and prises off the shell. But a limpet is a much tougher proposition. It has been said that a sudden kick from a boot can dislodge a limpet from its rock-hold, though if it has previously been gently tapped and induced to pull its shell down firmly against the surface of the rock, it cannot then be dislodged by the most brutal kick, notwithstanding that the surface of its rock-hold may be irregular, for the broad base to its shell permits the broadest area of attachment. Its conical shape also offers the minimum resistance to pounding waves; the rougher the seas, the more firmly does the limpet cling. Nevertheless, what boot or breaker cannot achieve by force, the relatively weak bill of a bird achieves by technical application, for by striking a limpet's basal edge a sharp angled blow with the tip of its beak, an oystercatcher can not only dislodge a small limpet, but shift larger ones slightly out of position and weaken their hold sufficiently for it to prise its bill under their shells and lever them off the rock.

An oystercatcher can also open bi-valve mussels up to 2 inches long, though the precise technique employed depends upon whether the mussels are under water or exposed on a scap at low water. If the mussels have been dried out for some time during the period of low water, they are tightly closed. In these circumstances, M. Norton-Griffiths and other ornithologists who have made a close study of the oyster-catcher's various techniques, have observed that it pulls a mussel off the scap and carries it to a patch of firm sand. There it places the mussel upside down, so that its flat ventral surface is exposed, and strikes the cleft between its valves with a rapid succession of as many as twenty-five oblique hammer-like blows, the number depending upon the size of the mussel. The ventral surface is not only very much weaker than the pointed dorsal surface, but also offers a more stable target. Under this treatment a chip of shell is ultimately broken off from one or other of the valve edges and a large semicircular hole bared, into which the oystercatcher can insert its compressed bill, prise the valves apart, and peck out the flesh. Cockles are opened in much the same manner, though since any portion of their shells is breakable, no special part is attacked.

Norton-Griffiths observed that a rather different technique is employed in dealing with mussels that are still covered by the tide, for the valves of these are slightly agape to permit the water to flow in with its abundant microscopic organisms. In these conditions the oystercatcher, wading slowly through the shallows, has two options. It can strike its beak sharply into the cleft between the valves and, jarring one out of alignment with the other, engineer a gap into which it can insert its beak. Alternatively, it can stab through the gaping valves with a single hard thrust and cut the large abductor muscles that hold the valves together. The mussel can then be prised open, the fleshy attachments along the edges of the two valves cut by a chiselling action (while at the same time, the beak, opening and closing very rapidly, works like a pair of scissors) and the whole operation completed on a 1½-inch mussel in only twenty seconds.

Northumbrian mudflats also provide food for thousands of duck and geese. Pied shelduck, plodding steadily over the mud 50 yards one way and then 50 yards back again, superimpose a maze of web-prints over the marks made by their beaks, as they push and dredge through the ooze with them, snake them from side to side, and sweep them to and fro in one place in their quest for mollusks and crabs and small fish. Wigeon and brent geese, after resting in immense rafts on the sea during the period of high water, flight in their thousands to the raised swads on the slakes when the tide begins to recede, to feed on the succulent white roots of the eel- or crab-grass, the *Zostera marina*, whose band-like leaves undulate in the ebbing waters. And there are the eider-duck, more than half of whose food is mollusks and a quarter crabs. One tends to associate eiders with rocky coasts and islands, from whose cliffs one can look down and see them slipping under the waves to descend 10 or 20 or 35 feet into the translucent green depths, where the sinuous forms of seals sport; down to the brown rocks and tangle on the bottom, over which they row with their wings, before inclining upwards, wings now bound tightly to their sides, after being submerged for a quarter of a minute or even a full minute. There is also the familiar spectacle, when storm waves are rolling into the coves in an endless succession, of eiders tossing on the swell and battling with the greatest determination against the pull of the undertow down the steeply sloping shore, while imbibing quantities of small organisms that also provide a feast for gulls and every kind of wader. But those fleets of eiders whose habitat includes mudflats pass much of their time up the creeks among the mussel scaps at all seasons of the year, except when the ducks are nesting on the islands, though as soon as the ducklings have hatched they are ferried across the sea to the quiet backwaters of the slakes. At intervals the eiders go ashore to rest for long periods on the sandbanks protruding into the mud-flats, depositing on these spits hundreds of small pyramids of blue mussel shells, especially those of the ½-inch young ones, which they swallow whole, and of which a single eider's crop and stomach may contain as many as 150; but they also hunt for the green shore-crabs, which may measure more than 2 inches across the carapace. Not only do they dive over the reefs for these, but also up-end in the creeks like any farmyard duck. Bringing a crab to the surface, the eider literally shakes its claws off, often by holding it by one claw and beating its body on the water, in between efforts to swallow it whole; but even when it has been de-clawed this devouring is a difficult operation, and the crab is usually dropped several times, though retrieved before it can sink to any depth.

The salt tides of the sea penetrate many miles up the estuaries to enrich the herbage of the salt marsh peninsulas that protrude into her domain. Raised 2 or 3 feet above the flanking mudflats or sands, the salt marshes may be covered by the sea only a few times a year, though every spring-tide will flood up the deep creeks that wind far into them. Some marshes, indeed, are totally submerged only once in several years when a special combination of spring-tide and gale-force winds results in a high bore, a miniature tidal-wave, storming up the estuary on a broad front and sweeping over the marsh with such impetus that the sheep grazing on it are trapped and drowned. Fertilized by these periodic deposits of tidal salts and manured and kept in good order by sheep and cattle during the summer months and by geese during the winter, a salt marsh provides unrivalled grazing. Barnicle geese and graylags crop the grass almost as closely as that of a bowling green or tennis court, for both of which the turf from one marsh in the Solway Firth was exported in days gone by; while the short-billed pinkfeet geese pluck up the plantain weeds, pitting the marsh with deep holes as if a herd of small pigs had been rootling. In the winter twilight of late afternoon, tens of thousands of wigeon, mallard and teal, together with a few small waders and golden plover, fly into a Solway marsh from the adjacent sands and mudflats to feed during the night around the pools and in boggy depressions where creeks have overflowed. And when a bore is driving up the estuary and flooding over the flats, flock after flock of oystercatchers gather on a sandy spit of the marsh. From south-west to north the sky is etched with long lines of oystercatchers, strung out right across the horizon and converging on this spit from every quarter of the great estuary. In many skeins there are 1,000 or 2,000 birds. For an hour they come up in their wavering formations, to mass on the spit, until perhaps 20,000 carpet the sand. And as the tide swirls higher and higher the thousands at the seaward edge keep flying back over their fellows, until their black carpet stretches away beyond the eye's discerning and is lost in the haze. Once the half of them rises as one, with a thunderous roar of wings. And what a piping there is, while they stand about idly or bathe in pools—until the waters begin to ebb, when there is absolute silence, as one small group after another takes off for its feeding grounds.

16: Sea-Urchins and Sea Otters

That narrow band around islands and along the coasts of continents where the land confronts the ocean possesses its own specialized fauna, though the composition of this varies according to whether the land frontier is rock or mud or sand, and depending upon the relative force of the breakers and the extent of the inter-tidal area. The small fauna—shellfish, worms, crabs—inhabiting the reefs, sandy shores and mudflats of the between-tides zone are subjected to drastic environmental changes every day by the ebb and flow of the tides. During one period of hours their habitat is immersed in comparatively cool salt water subject to only slight fluctuations in temperature; but during the succeeding period it may be drying out under a hot sun or possibly drenched by heavy downpours of fresh water.

The sea's influence may, however, reach beyond the high-water level of normal tides into the spray zone, though this is wetted only by storm waves and during spring-tides, and cliff faces may be dry for days at a time and bared to the heat of the summer sun. Nevertheless, in crevices in the rocks or on the piles of jetties, small gastropod snails and barnacles have established their niches in this unstable habitat. During a dry period they are to all appearances dead. But engineer a flow of water over them, and the barnacles open the valves of their shells. Six pairs of feathered feet or *cirri* spread out like a net and then, fully expanded and curving inwards to prevent the escape of any planktonic minuti they may have ensnared, are quickly withdrawn with a clutching motion. So long as the water flows, the barnacles continue to feed rhythmically in this manner, "kicking" their food into their mouths; but cut off the flow, and they retract their *cirri* and bar their shells with, in Klingel's words, two solid plates of ivory which fit so tightly that the shell is both airtight and watertight. Barnacles have not perhaps fully exploited their capabilities, for they display remarable resistance to drought under laboratory conditions. Witness the hundred acorn-barnacles which an Italian biologist kept in a dry state on his desk. Although these were placed in sea water for only a day or two at intervals of two or three months, and were in fact watered on only 59 out of a total of 1,036 days, their yearly mortality was no higher than 10 or 12 per cent.

Perhaps the barnacles have in fact established themselves in a less exacting environment than that of the permanent inter-tidal zone, whose inhabitants are subjected to hourly variations of humidity and temperature, and to varying pressures from pounding waves and seething surf. In his account of an Inaguan reef, Klingel has described how "Day in and day out great walls of surging water flung themselves at the cliffs, dashed high in the air and fell back again; high on top of the cliffs great blocks of coral and sandstone weighing many tons were piled in a long rampart where they had been deposited by the ocean during storms."

There are two ways of surviving the impact of such terrific forces, equivalent to 25 tons on a square yard. Either you give way before it, like the palm tree bending to the hurricane, or you equip yourself with armour and cement yourself to the rock. Klingel has demonstrated these alternatives:

> Just before my eyes was a cluster of delicate hydroids, flower-like animals with tentacles so diaphanous and translucent as to be illusory. Near by hung long tendrils of filmy algae . . . tiny interwoven threads that parted at the touch. How could they survive the tons of surf that poured on them every few seconds? Simply by giving way before it, by swaying in the direction of the water. . . . The anemones and lacy algaes . . . persisted because they eternally agreed with the superior force confronting them. . . . The chitons or coat-of-mail shells, the limpets and gastropods . . . moved not a single inch; clad in impenetrable armour, imprisoned behind walls of ivory or unyielding calcium, they sat immobile; neither force of wave, heat of sun nor attack of enemy caused them to change their mode of living one iota. . . . Those individual limpets which lived at the places of the most violent action had the thickest shells. . . . The mussels . . . used a system of resistance unlike any other creatures of the surf. By some marvellous chemistry they derived a substance from the sea water which they spun in long silken ropes which they cast out in all directions . . . anchoring them firmly to the rocks.

Actually a mussel's anchor lines, or byssus, originate in a thick fluid running along a groove, which extends down the hinder surface of the thin but very extensive foot. Where the fluid meets the rock at different points it spreads out into discs and almost instantly hardens into the tough threads. Thus the mussel is, in the words of C. M. Yonge, guyed to the rock by a divergent mass of threads, a proportion of which can take the strain of wave impact from whatever direction this may come. If any threads are broken, others can be formed. Nevertheless, the surf is sometimes too rough for a colony of mussels and, with the entangled byssuses all giving way at once, the mussels are precipitated into the sea *en masse.*

But perhaps the most remarkable inhabitants of this turbulent region are the sea-urchins, which are able to withstand the impact of the surf through the suction powers of hundreds of tube-feet—as many as 2,000 in some species. These are arranged mainly in ten symmetrical double-rows radiating from the centrally located mouth on the underside of the urchin's barbed, almost spherical test or casing, which is supported by five tightly fitting skeletal plates; but are also scattered over the test's upper surface. Since these feet are capable of contracting and expanding as much as 2 inches, they grip the surface of the rock firmly at all points, no matter how uneven it may be, and cannot be dislodged by the surf. Some species have evolved another technique for retaining their niches in the surf zone of coral reefs, for by secreting a powerful corrosive they etch out burrows in the soft coral. These borings are completed at an early stage in their lives, and though they can continue to corrode their interiors as they grow, their bodies become larger than the original entrance to the burrow, and they become prisoners in the coral. Although the burrows are reported to be invariably grazed bare of all algae growth, their occupants must presumably be able to extract sufficient plankton and fragments of algae from the water flowing past their caves. By what means they would capture these is not altogether clear—though possibly they are able to extend some of their tube-feet as the heart-urchin does from its burrow in the sand—for the normal free-moving urchin grazes on algae and seaweed, grinding these up in its peculiar five-sided jaws. However some food is also obtained fortuitously in a most ingenious way described by Klingel. He observed that though at times the surf piled loose carpets of torn seaweed over the boulders and cast up gritty piles of gravel and coarse sand on his Inaguan reef, the urchins were always spotless; no grains of sand, strands of algae or blotches of parasites marred their jet-black tests:

> When a fragment of sand or dirt fell between the needles, it was grasped by a tiny clamp or pincers equipped with a triple set of jaws, like those of certain types of dredges; these were mounted on a flexible shaft of muscle and skin which transported the debris to the next claw which carried it to another or to one of the tube-feet . . . and so on until the offending fragment was dropped into the water. Small parasites that slipped between the barrier of spines were not treated so gently. The moment they touched the urchin's side the claws began snapping, opening and shutting until they seized on some portion of the animal's anatomy. Once a pincers secured a grip it held on tenaciously and then if a struggle ensued, other claws came to its rescue until the captive was held rigid by myriads of tiny clamps. Only death of the parasite caused the grip to be relaxed; the corpse was then passed from claw to claw, foot to foot until it reached the urchin's mouth.

Every portion of a sea-urchin's test is armoured with spines, totalling several thousand and mounted in ball and socket joints. In some tropical species the 12-inch-long spines are covered with a poisonous mucus and, being brittle, break off where the barbed point enters the flesh. Nevertheless, just as some sea-anemones shelter small fish within their poisonous tentacles so, on coral reefs, a small blue fish lives within the sea-urchin's armament of spines.

With the aid of their tube-feet, and also that of their mouths, the urchins can move about slowly in series of wavy rhythms, though some species make use of their spines for locomotion over sandy bottoms. Off the east coast of Madagascar, schools of several different kinds of urchins, accompanied by predatory mollusks and starfish, have been observed migrating in hundreds at night at the very respectable rate of 450 yards an hour, possibly when in search of fresh feeding grounds; and off the Pacific coast of America enormous armies of them have in recent years seriously depleted vast areas of the giant kelp and the brown or bull kelp, which stretch parallel to the coast from Alaska to Baja California, and which are also being adversely affected by pollution and by high water-temperatures.

The kelp introduces us to another habitat at the sea's edge, characterized by strong currents and heavy swell, whose stress it is peculiarly adapted to withstand. The kelp is anchored to a solid bottom, usually of rock, by a massive tangle of grappling roots, which may be more than 5 feet in diameter, and which serve as a base for as many as fifty stems or stipes, but do not provide the plant with any

Kelp sea-floor group.
From the top: painted greenling, yellowfin fringehead, sea urchins, and blue-banded goby

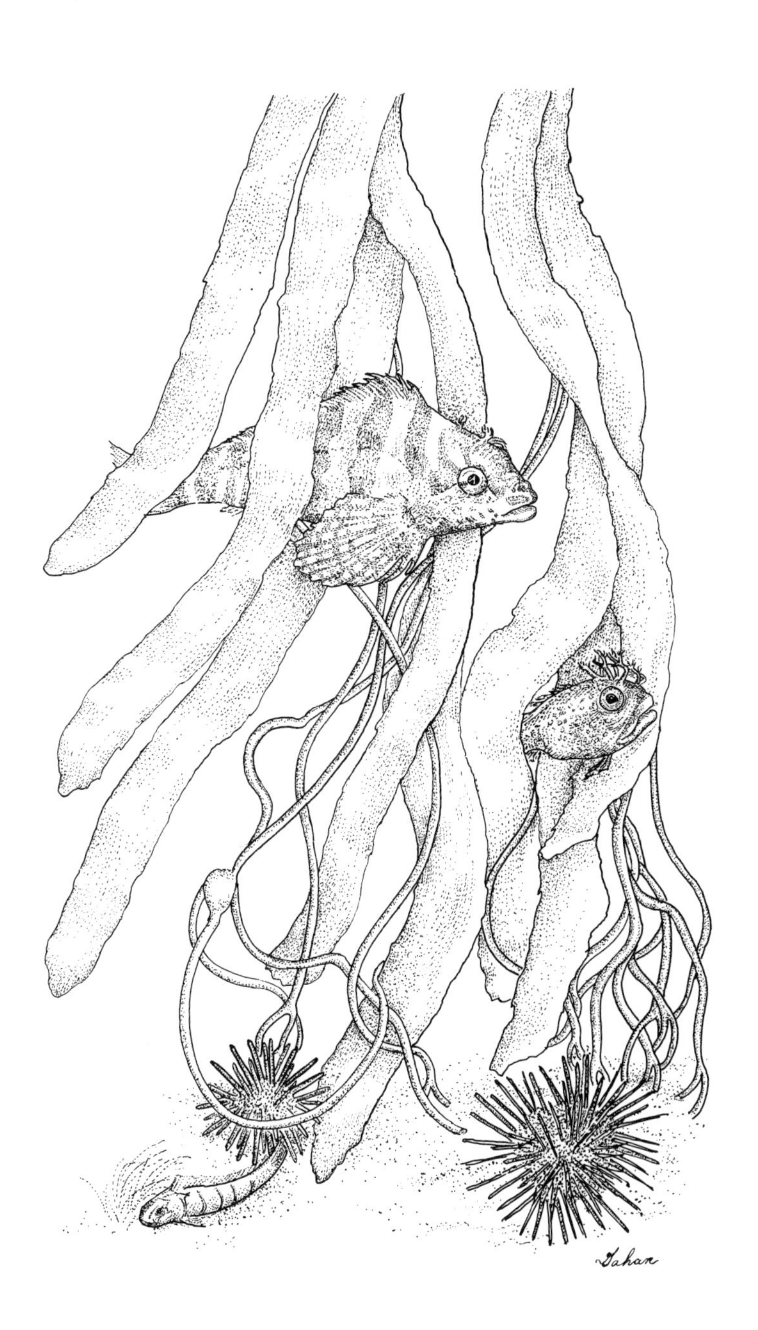
Gahan

nourishment. Edmund S. Hobson of the University of California has described how off the coast of California, where the giant kelp is anchored in depths of 100 feet with stipes (stalks) growing to lengths of 160 feet, majestic columns of intertwining fronds reach up to the surface where, becoming entangled with the fronds of adjacent columns, they form a dense canopy. In areas where small breaks in the dense foliage allow narrow shafts of sunlight to penetrate down into the twilight, an almost cathedral-like atmosphere is created. Thousands of minute crustaceans, mollusks and tube-worms live within the tangled mass of the kelp's anchor, and from the sunlit waters of the canopy to the rocky shadows of the ocean floor the kelp forest supports a thriving population of predatory fish, matching in colour that of the kelp. Kelp bass and olive rockfish stalk smaller fish and crustaceans on the bottom; larger, 2-foot-long kelp-fish hover in a vertical position, aligned beside a column of similarly coloured kelp or stalk to within a few inches of their crab or fish prey.

The slender stem of each of the giant kelp's intertwined fronds produces large leaf-like blades along its entire length, with an oval, gas-filled supporting float at the base of each blade. The thicker but rather shorter stipes of the brown kelp terminate in large rounded floats with streaming apical fronds. At a distance these floats are indistinguishable from the bobbing heads of sea otters, rising and sinking in the swell. The otters are the pride of the Californian kelp fields, for fifty years ago they were presumed to be extinct off the Pacific coast of America, with the exception of a small herd at the Aleutian Islands, while a few scores were known to have survived at the Kurile Islands, the Commander Islands, and Bering Island in the Bering Sea. Their range originally extended from Baja California north-about to the

Kuriles, and thus included both arctic and tropic regions. But they were ruthlessly slaughtered by eighteenth-century fur hunters, and within twenty years of Bering's initial voyage of exploration in 1741–2 the Commander Island's otters had been virtually wiped out, though they do not appear ever to have been very numerous, for the Russians' average annual kill of this easily hunted animal was only 1,500 during this short period. The survivors retreated to the Kuriles and the Aleutians, reappearing at the Commander Islands when the otters of those archipelagoes were also exploited, after the Russians had discovered the Pribilof Islands in 1786. Indeed, it was the otters, and not gold, that initially brought California to the notice of the Russians, British and Spanish, and it was the otter-skin trade that was responsible for the deepest penetration of America by the Russians to Fort Ross (a few miles north of Russian River in Sonoma County), which was the headquarters of their Aleut slave hunters. Remnants of the other herds survived, however, in the wildest and most rugged retreats up and down the Pacific shores of America, including—though this was not realized until 1938—the coast of California south of Monterey, whose fringing rocks and islets are also the haunts of sea lions and fishing flocks of cormorants and pelicans. International protection had been afforded the otters as early as the years 1911–13, but their initial recovery was very slow, since the single kit is gestated for eight or ten months, and subsequently remains with its mother for a further fifteen months; thus the female breeds only once in two or three years, and in the average herd adults outnumber young by a dozen to one. However, after forty years' official protection the otters' numbers suddenly increased sharply in all areas during the 1950s. By 1956 their population of between 4,000 and 5,000 along the 120-miles coastline of Amchitka Island in the Aleutians had possibly reached saturation point, so far as food sources are concerned, for there is evidence of considerable mortality in hard winters. By 1970 the Monterey herd of 94 had increased to 1,200 or 1,500 and was extending its range south to Moro Bay, and the world population of otters was estimated at 36,000.

The sea otter—4 or 5 feet in length and up to 90 pounds in weight, or 65 pounds if a female and from 3 to 5 pounds if a kit—is a large relative of the river otter and, as such, a member of the weasel family, the *Mustelidae*. It is essentially marine in its habits, but not oceanic, though one has been encountered 60 miles offshore. It is ill-adapted to life ashore,

Sea otter with kit; kelp in foreground

on which its long supple body, which has been likened to a liquid-filled bag, waddles awkwardly and heavily with apparently great exertion, and seems barely supported by its legs and flipper-like hind feet, though if alarmed or threatened, it can hump along quite rapidly and even jump clumsily. In the sea, by contrast, the otters enjoy perfect mastery, swimming swiftly on their backs while paddling with their webbed hind feet. They might therefore be considered intermediate between the seals, swimming with flippers, and the river otters which propel themselves with their tails. When travelling at speed, however, during play or if alarmed, they usually swim in the conventional posture, with short forearms folded across their chests, propelled by powerful strokes of their hind feet striking in unison or alternately, and glide and undulate smoothly through the water with the sinuous motion of an eel.

A naturalist, acquainted only with the Californian race of sea otters, might indeed suppose that they were exclusively marine in their habits, for these otters pass their entire lives of perhaps fifteen or twenty years in the kelp fields a few hundred yards offshore. There they mate, bear their kits, feed, play and spend a great deal of time sprucing their fur. Throughout much of every calm day they rock gently on the waves. Some sprawl on their backs, with hind feet and relatively long, rounded and flattened tails sticking straight up out of the water; others sleep, with forepaws folded on their chests or shading their eyes from the sun. It is even reported that at night they anchor themselves to the kelp, either by hooking their paws over a stipe or by wrapping a few strands around their bodies, in order not to drift ashore or to lose contact with other members of their herd. Although this seems to be a well established practice, Russian zoologists do not accept that it is a deliberate action, contending that the otters merely become entangled in the strands while playing in the extensive submarine thickets of sea-kale which, in the Bering Sea, replace the fields of kelp.

These northern otters, unlike those off California, have not entirely abandoned a terrestrial existence in their habitat of rocky peninsulas, offshore reefs and small rocky surf-drenched islands. On the Aleutians they often haul out to sleep on the rocks on calm summer nights when tides are low, though seldom venturing more than a few feet from the edge of the sea, while the females frequently bring their kits ashore about sundown and remain on the rocks until dawn; but they are on shore in greatest numbers when storm waves make it difficult for them to dive for food. On the Commander Islands, however, they constantly haul out to rest by day on large isolated rocks and also on reefs and beaches not more than a dozen yards from the sea, and regularly sleep at night on land, even when it is snow-covered. Indeed, the female apparently gives birth on land and, if the rock has a protective mat of seaweed, may lie on her back with the kit on her chest, as she does in the sea, holding it in her paws, hugging it, lifting it into the air, and flipping it from side to side in very human fashion. She may leave the kit on a rock when she goes fishing or, alternatively, when she climbs down from the rock into the sea it may hang on to her with its teeth, though apparently still sleeping, swaying back and forth like a sack and bumping against the rocks. This behaviour contrasts with that of the Californian female, for when the latter dives for food, which rarely takes her longer than a minute, the kit, previously sleeping and suckling on her chest, now sleeps on the water, buoyed up by the air trapped in its waterproof coat of long hair, constantly groomed by its mother's sensitive mobile paws. Although the kit is suckled for so many months this does not inhibit it from begging for and receiving portions of sea-urchin or other mollusk when its mother is feeding; and eventually breaking open an urchin itself by pressing the shell with its paws and scooping out the soft parts with its tongue and lower incisors.

It has been suggested that the Californian race of otters is now exclusively marine as a result of persecution, but since those on other Pacific coasts have been no less persecuted, but still go ashore, this is obviously not the case. More probably it is a matter of environment. In contrast to the kelp fields, which hold abundant supplies of otter food, there is little food in the onshore sea-kale thickets, though those further offshore grow over banks or shoals where the bottom is accessible to the otters. The main attraction of the sea-kale would appear to be that its thickets tend to provide calmer patches of water, though since the kale's surface areas are destroyed by autumnal storms, the otters are obliged to remain close inshore during rough winter seas, until the kale grows again in the spring. Perhaps the kale thickets also serve as a refuge from killer whales, for the otters are reported to come close inshore when large packs of killers appear off the Commander Islands.

In all parts of their range 60 per cent of the sea otters' food is reported to consist of sea-urchins, and since a herd of otters has been known to clean out all the urchins in its feeding area, their increasing numbers in Californian waters should assist in reducing the hordes of urchins currently destroying the kelp by cutting it away at the root. The proportions of other foods taken by otters vary according to locality. Off California 35 per cent consists of common black mussels, but only 5 per cent consists of that shellfish with which sea otters are popularly associated, the abalone. Off the Commander Islands, however, mussels are only occasionally eaten, though the Russian zoologist, Barabash-Nikiforov has described how an otter, when feeding littorally "Swims along the margin of a rock, carefully examining the stands of seaweed covering the slope. . . . When it finds a clump of mussels, it turns on its flank, and begins to energetically thrash at it with its forepaws, until the byssuses fixing the clusters are torn or pulled free; it then easily opens the shells with its teeth and bolts the molluscs."

In addition to enormous numbers of two kinds of sea-urchins, especially a shorter-spined species which concentrates in densities of 200 to the square yard in depressions in the reefs that hold water when the breakers recede from the 15-to-30 feet zone, the Bering Sea otters' diet is made up of about 17 per cent crabs and other crustaceans and 23 per cent mollusks, which are reported to have abnormally thin shells, because of the low calcium content of the Bering Sea waters. But these percentages do not take account of the large number of octopus eaten, in addition to such fish as cod and the Pacific lump-fish; for a male otter requires 15 pounds of food a day, and a female from 8 to 10 pounds.

The otters feed most actively before dawn and in the early morning, when the herd splits up and individuals scatter in all directions. After this first feeding session they rest and groom themselves, for which purpose those in the Bering Sea may clamber out on to the reefs or go ashore. At midday they feed again, then rest, groom and perhaps mate until the day's final feed between early evening and nightfall, after which the herd assembles at a kelp bed and settles down for the night in a compact mass, though if it is moonlight some individuals may continue active. The bulk of the otters' food is obtained during a 60-second dive at depths of from 5 to 60 feet and, though reported to be able to dive to 300 feet and remain submerged for 5 or 6 minutes, they rarely dive deeper than 100 feet. Diving obliquely, rather than straight down, an otter swims at an angle to the bottom, perhaps combing it for shellfish with its tactile whiskers; then, with the aid of its forepaws which, though stubby and rounded, terminate in small cushions enclosing short but mobile fingers adapted to seizing and holding, it scoops up five or six or as many as ten urchins, or perhaps a mixed bag of six urchins and three rock oysters, and clutching these to its chest between one foreleg and a fold of loose skin, which forms a pouch on either side of the mid-line of its chest, swims up to the surface with them. There, rolling over on to its back and tucking all but one of the mollusks into the skin pouches, it crushes the spines of the urchin it has retained in its paws and also presses the armour around the mouth part. Then, having gnawed along the line of fracture or having bitten a hole in the test, it licks or sucks out the fragments from the lower part of the shell, together with the contents of the intact upper half, or scrapes the soft parts into its mouth with its paws—rolling over every half minute or so while eating in order to cleanse its fur. The oyster it pries open with its canine teeth and separates the valves by twisting its paws sharply; then, dropping the empty half on its chest, holds the full half in both paws and licks out the orange and white body mass. Having cleaned out one mollusk, the otter may then indulge in its favourite play of twirling round in the water with a propellor-like motion, while clutching its pouches with its paws so that none of the mollusks fall out, and deliberately splashing water over any gull that ventures too close to pick up a stray morsel; for the whereabouts of a herd of feeding otters is often revealed by a cloud of gulls circling overhead.

Octopus, fish, and crabs are also eaten while the otter is floating on its back, and if the fish is a large one, such as a cod, the otter first gnaws through the spinal column and tears away strips of the back muscles, before eating the tail, the internal organs and finally the head. Large crabs are placed on the otter's chest and their legs pulled off before their shells are cracked with the otter's flat cheek-teeth and the meat disposed of in five minutes. But the Californian otters have evolved a most remarkable elaboration of this feeding technique, for in dealing with certain hard-shelled mollusks an otter of this race fetches up from the bottom with its catch of shellfish a large stone or piece of rock from 2½ to 6 inches in diameter and weighing as much as 8 pounds. Though it presumably makes use of its pouches in transporting this load to the surface, one wonders how it is able to collect both shellfish and rock in a single operation. However, on surfacing, the otter rolls over on to its back in the usual way and, placing the rock on its chest, uses it as an anvil on which to hammer the mollusk, holding the latter in both paws and bringing it down hard on the rock with a full arm action from well above its head. Although the otter's chest is said to be cushioned with a special roll of fat, serving as a shock-absorber, the click of the impact can be heard at a distance above the noise of the breakers, and a rapid tattoo of several vigorous blows may be required before the mollusk's shell is cracked and its soft parts bared. An otter may sometimes carry its anvil with it throughout a succession of food-collecting dives, and one was observed to open no fewer than 54 mussels by striking them on its anvil 2,237 times in the course of 1½ hours. That this employment of a tool is not merely a stereotyped action but intelligent has been demonstrated by Karl Kenyon, whose captive Alaskan otter caused him much trouble by repeatedly removing the mesh strainer over the drain in her pool and, when this was secured by and iron band and bolt, by pounding the bolt with a rock until she had achieved her aim. Although the use of an anvil has not been reported among wild Bering Sea otters, the fact that the hair on the chests of some individuals is abraded suggests that rocks may be used for this purpose, and Kenyon's captive smashed clams with her stone.

Californian otters are often portrayed opening abelones in this way. Seven kinds of abalones inhabit inter-tidal and shallow waters along the Pacific coast of America; the largest, and perhaps the most favoured by the otters, is the red abalone, which may be as much as 8¾ inches long and 6¾ inches wide and several pounds in weight. But in fact an abalone is not, and need not be, broken open because it is a univalve with an oval, very flattened, coiled shell pierced by a characteristic series of openings; its soft underside is therefore exposed when the otter very quickly prises it from its rock-hold. How the otter does this is not known. Possibly it selects those abalones that have been weakened by boring worms and mollusks; possibly it pounds the abalone with a stone or bites a large piece out of the shell with its strong canine teeth, for all abalones that have been opened by otters are distinguished by a large hole from which a piece of shell has been broken off. But whatever technique an otter employs in obtaining its abalone, it does not open it on an anvil, but bites large portions out of it while holding it up in its paws; and if a portion is so large as to necessitate the otter using its paws to hold the food to its mouth, the remainder is placed on the chest or abdomen, from which it does not roll off during the five minutes required for its consumption, even when the otter is awash in rough seas.

No doubt the Californian otters included a higher percentage of abalones in their diet at one time, before exploitation by the Japanese and Chinese (who fished them both for their meat and for their mother-of-pearl) resulted in abalones becoming relatively rare, so that they are now, like the otters themselves, conserved only by rigorous protection; for despite their dramatic recovery the otters' future is not entirely assured. There has been a recent reduction in the strength of the Californian herd due to illegal shooting by fishermen, believing erroneously that the otters are responsible for the decrease in the numbers of abalones; while the 1971 nuclear test on Amchitka killed several hundred otters, and must presumably have affected both the environment and the food sources of those that survived. No doubt there will be further nuclear tests in the Aleutians, and there is the no less serious threat of large-scale pollution from the Alaskan oil-fields. Oil slicks could prove particularly lethal to the otters because, unlike other marine mammals, they are not equipped with a layer of blubber with which to retain body heat, but rely on a jacket of air trapped between the soft, fine and silky hairs of their dense under-fur. The latter also protects those inhabiting arctic waters from freezing-up, for their outer coats, in contrast to those of river otters or fur seals, afford little protection. Thus, if oil penetrates the sea otter's closely packed several hundred million fibres of under-fur, its thermal insulation is destroyed. Hence its incessant grooming.

17: Where Seals Haul Out

On a warm October day, with a fog bank a mile or two offshore, there is no pleasanter place to sit than 300 yards out at the seaward end of a jutting North Sea reef when the sea has ebbed right back from it after a high spring-tide. On either side are submarine forests of whip-like thong-weed and heavy tangles of oarweed with broad orange-brown fronds and long thick stems, now submerged, now awash, swaying and seething to and fro in the smooth sea-swell. It is a strange but wonderfully peaceful primeval world, belonging to the great gray seals—the Atlantic seal—seven or eight of whom are continually bobbing up their heads inquisitively to gaze fixedly at me, whether I am statuesque in the clammy corner of two dripping rock walls or sitting on the knife-edge of a limpet-studded reef. I am right out in their element, with their ocean rolling past on either side, seeking to suck me down from my slippery niche, and close enough to them to see the bloodshot rims to their dark eyes. They like to stand, five or six at one time, in the sliding flank of a wave, riding the swell without apparent loss of position, while singing and sighing *boo-oo-oo-oo-oo-oo-oo*—a weird but lovely sound, sad and haunting. And then, elevating bristled muzzles, they close their cavernous eyes, open and shut the dark cavities of their broad nostrils, and slowly sink beneath the waves.

Although the sea is their element, in which they dive with the grace and agility of shags or shoot through the wave tops with the sinuous swiftness of salmon, leaping clear of the breakers with a tense curving of smooth gleaming bodies marbled in black and olive, they are drawn to the land and the heat of the sun; though it is true that there are never more seals lying out on the rocks, wailing and keening, than on a calm foggy day. However, on a June day, when the wavelets lapping the base of Gannets Rock on Lundy Isle in the Bristol Channel begin to ebb, the seals cruise lazily through the surface waters of the pellucid green sea. Their glistening hog-backs, the colour of old pewter or of smooth shining rocks, are humped like those of porpoises when they bore down into the opaque underworld, where their luminous, glaucous forms, swirling away from one another, set the small auks above pattering over the waves. When the sea has ebbed from the base of the Rock the seals seek the sun, heaving themselves up on to the broad ledges with the aid of the strong claws on their fore-flippers and pressure from their hind flippers, often attempting to scale the slippery rock face long after the waters have ebbed too far back for them to leap up its steep wall. There is great gripping strength in their fore-flippers which are prehensile with distinct fingers, though they cannot of course be used separately but only as a single instrument. Maintaining a

Gray (or common) seal cow

hold with these, their belly muscles contract and expand as they heave themselves upward and forward.

Those that have been able to haul out, roll over on to sides or backs, recline at full stretch, and expose their rotund bellies to the glorious warmth of the sun. From time to time they heave themselves higher up the bare rock—their drying backs now the colour of gun-metal with black flakes—flounder about sluggishly, coil and uncoil, stretch, scratch their bellies lazily, wipe their muzzles with their flippers, and yawn continually, revealing pink throats. Highest up on the Rock bask the fox-coloured yearlings, pale dappled brown with no black blotches on throats or bellies. Although their mothers deserted them when they were less than a month old, the cows object to the bulls hauling out near them. One cow, indeed, rolls over at a bull which is clinging, with head and shoulders out of the water, to the shelving rock, snapping at him with gaping pink jaws, while he turns his massive anvil-shaped head from side to side, loath to release his hold; and the two growl and snarl like dogs until the bull, moaning irritably, finally plunges away and subsequently rises, puffing from the swell. Later, he attempts another landing and the cow, at full stretch, wakes up with a sudden start, snorting angrily. When he is ultimately allowed on to the Rock he moans at the yearling sulkily, cuffing it with his flippers.

When north-easterly gales roll up big seas around the Rock the seals favour the flat reefs that extend 200 yards out from the cliffs on the Atlantic coast of the island; and here fifteen fat "slugs" will lie out at one time on different reefs. Most are gray and white, some brownish-black or stone-yellow or slate-coloured, and one is a furry, almost sable, black. When the tide begins to flow and surges over the reefs they are reluctant to leave their sun-traps, lifting their whiskered heads petulantly from the dashing spray; and for 20 or 30 minutes thereafter obstinately and repeatedly raise their chests and sterns high out of the swirling water. In the end, however, they are too wet to enjoy the sun any longer and, with moanings and roarings, are washed off the reefs by the waves. For a while after, they play in the water, turning over on their backs and striking at one another with their flippers. Their speed under water is prodigious, and at a distance of 150 yards they might be torpedoes.

Some marine mammals, such as whales and dolphins, are totally divorced from the land. Even the Californian sea otters are, as we have seen, apparently marine and actually bear their pups at sea. But, with the exception of the walrus, all seals normally haul out on to land or ice to pup, and all, including the walrus, like to haul out to sleep or bask in the sun, for all must rest after their activities in the water. They can of course sleep at sea. Gray seals are often to be seen sleeping in the sea. A bull may do so in a prone position at the surface or under water, though more commonly in an upright posture, with most of his head above water and his muzzle pointing skywards, for as long as half an hour without opening his eyes. Ronald Lockley, to whom we owe a great deal of our knowledge of both gray and common seals, has described watching two gray cows sleeping on the bottom in the shallow waters off a Welsh beach. At intervals of from 5 to 9 minutes they would rise to breathe ten or fifteen times, and were probably sound asleep throughout this operation, for they usually surfaced without opening their eyes and then somersaulted automatically in order to dive down again head first. He conjectured that they might sleep on the bottom during storms, since they can remain submerged for at least ten minutes and probably for twenty minutes.

For the purposes of pupping and mating, numerous herds of fur seals and elephant seals travel hundreds of thousands of miles to traditional island breeding stations, on which they concentrate in rookeries of thousands or tens of thousands; but sea lions, gray seals and common seals gather in much smaller numbers, though there are qualified exceptions. Two thousand Steller's sea lions of the Alaska race have, for example, established a rookery on Ano Nuero, an island off central California, at the extreme south of their range; while the 200-strong herds of southern sea lions, which haul out on the often widely separated islets of tussock-grass off the Falkland Islands, congregate during the breeding season in colonies of as many as 3,000 on adjacent islands and coasts where strong tide-rips produce upwellings of plankton. These attract the fish of the extensive kelp fields, and thus provide the sea lions with an easily accessible food supply. The latter also patrol the routes followed by penguins commuting between nesting colonies and fishing grounds; but though they capture numbers of penguins, eviscerating them with the leopard seal's technique, they perhaps do so for sport rather than for food, because the tide-line can be seen strewn with the intact corpses of penguins.

One of the problems attendant upon the immense concentrations of such large animals as seals in one place is obviously that of food supplies. At a guess, the 1½ million fur seals, gathering annually to breed on the Pribilof Islands and give birth to 600,000 pups, might consume about 6,500 tons of food daily, and even if the bulk of this is in the form of euphausian shrimps, one must doubt whether even the multitudes of shrimps could supply such demands over a period of months. Moreover, before their periodical persecution, beginning at the end of the eighteenth century, there were perhaps twice or thrice this number of fur seals on the Pribilofs. However, this problem is lessened by the fact that the bull fur seals and elephant seals, which are double or treble the size of the cows, do not feed (or drink for that matter) throughout their residence on the islands. When they arrive at the rookeries, in advance of the cows, they carry sufficient reserves of fat, particularly on neck and thorax, for them to remain on or near their territories, in the company of the scores or even hundreds of cows they may acquire as harems, for several weeks without ever going to sea to feed. Then, when the bulls have mated and the cows have pupped, the members of the rookery disperse again to numerous fishing grounds over a wide area of ocean, though the bulls of the northern race of sea elephants which breed in midwinter, return to the rookeries during the summer in order to moult. The northern race of fur seal cows, for example, whose pups are born in July, desert the latter in the late autumn and, accompanied by the immatures, disperse over the north Pacific, with some herds migrating as much as 3,000 miles to winter quarters off southern California: whereas the majority of the breeding bulls and bachelors appear to remain in Alaskan waters until the season of winter storms.

Sea lions, however, reverse this order of dispersal, for though the bulls and bachelors have all departed from the rookeries by early September, the cows and pups, and also some yearlings, remain in them until about January. This results in some years in an interesting situation on Ano Nuero which, after all the Alaskan bulls and bachelors have left, is occupied by several thousand bulls of the Californian race, come up from the south to haul out on this island at the northern limit of their range. However, the sea lions' breeding behaviour differs from that of the fur seals and elephant seals. The master bulls of these collect their harems partly by actual combat, partly by an aggressive display. An obvious display property is the elephant seal's "trunk"; its precise function has not been satisfactorily established, but under some form of stress, this proboscis can be inflated into a trunk more than 12 inches long. Elevating his head vertically, the bull then curves the trunk round until it is inside his mouth and snorts rapidly through his nostrils, producing a burst of short, resonant and metallic clapping sounds, audible above the roar of the surf at a distance of half a mile. Therefore, according to one observer, if a bull can roar louder than his rival, the latter retires defeated; and only if both produce a similar volume of sound does a fight take place. Thus the bull with the largest trunk obtains the largest harem. But another observer disagrees, claiming that although the trunk is fully inflated during fighting or display, it has no special function, and that even if damaged, the strength, volume and pitch of the roar is not affected—nor its owner's chances of becoming a beachmaster—and that the roar is in fact produced solely by a forced exhalation from the lungs and by a vibrating of the soft palate. Possibly, it is merely the physical expression of anger.

Sea lion bulls, on the other hand, do not haul out, establish territories and collect harems around them, but pass much of their time ceaselessly patrolling the waters off a beach on which a few cows have pups. One bull controls this loose harem, and though he may occasionally slip away for a few hours, no doubt to feed, re-establishes dominance over any other bull that may have taken his place temporarily; but with constant territorial aggression between the beachmaster and aspiring bachelors no bull remains in possession of a beach for longer than a month. It has been suggested that the bull's constant patrolling also serves to protect the pups against attacks by sharks; but while it is true that he will herd the pups inshore if they attempt to swim too far out, whether they are his own offspring or those belonging to a bull on a neighbouring beach, he acts in the same way in shark-free waters, and it seems more likely that he is actually responding to his harem-herding instinct.

It has also been suggested that sea lions no longer breed in immense rookeries because of incessant persecution; but such persecution has not altered the breeding habits of fur seals and sea elephants. The population of the northern race of fur seals on the Pribilof islands of St. Paul and St. George was, for example, reduced by American skin hunters from between 1 and 3 millions in 1869 to about 134,000 by 1909; but this forty years of incessant and large-scale slaughter did not cause the fur seals to abandon these islands. On the contrary, after protection was imposed in 1909, their numbers increased within twenty years by 1½ millions, and has remained stable ever since despite the annual controlled cull of 60,000 or 70,000 young bulls, and despite the fact that the cows do not begin to breed until they are five or six years old and bear only single pups. Packs of killer whales visit, or used to visit, the Pribilofs every spring and autumn. The latter season coincided with the fur seal pups' first venture into the sea, and Dallas Hanna saw two killers take three adults in three minutes. In the spring of 1917 a pack of killers frequented the seas between St. Paul and Sea Lion Rock for more than a week, killing or frightening away almost every seal approaching the rookeries, and even grounding on the beaches in their ravenous pursuit of them. But although the stomachs of two of these killers contained respectively eighteen and twenty-four seals, predators clearly have little effect on seal populations in temperate seas. When however, entire populations are concentrated in these immense rookeries they are totally vulnerable to man the hunter; whereas their vulnerability would be much lessened if they bred in smaller dispersed groups. It is therefore interesting to note that in the British Isles, which hold about 70 per cent of the world population of some 50,000 gray seals, concentrated on either side of the north Atlantic, though overwhelmingly so on the eastern side, there are only two rookeries whose strength much exceeds 1,000; and since the breeding habits of gray seals are very different from those of fur seals and elephant seals, these must be termed colonies rather than rookeries. The largest colony is on the small oceanic island of North Rona, 43 miles from the most northerly point of the Scottish mainland and 45 miles from that of the island of Lewis in the Outer Hebrides. Its nearest neighbour is the still smaller island of Sula Sgeir 12 miles to the west. F. Fraser Darling, who with his wife and child was enterprising enough to winter on Rona in 1938, estimated that 5,000 gray seals were hauled out on the island during November and December, though they are now reported to number 8,000 or 9,000 and produce some 2,500 pups annually. (In calculating the strength of a gray seal colony it is assumed that the ratio of animals one-year-old and upwards to pups is 3·75 to 1.) Although the island has, significantly, been known as *Ron-ay* or seal island for at least twelve centuries, it has not been permanently inhabited by man since 1844, though two Lewis men wintered and died there in 1884–5, and since it is impossible to land from a boat on most days of the year, the seals are normally able to haul out on the island in perfect security.

Consider, in contrast, the location of the only other large British colony—the Farne Islands. The Farnes comprise from fifteen to twenty-eight closely-knit small islands and rocks, depending upon the height of tide and sea, with the most landward only 1½ miles from the mainland, and are therefore of relatively easy access in moderate weather. They are known to have been occupied by seals at least as far back as the fourteenth century, and there has been a long history of persecution by the local fishermen. In the past the colony does not appear ever to have numbered more than a few hundred; but with varying degrees of protection since 1920, the population has increased dramatically to more than

7,000, including nearly 2,000 cows, almost doubling in size between 1940 and 1950, while the annual crop of pups has increased from 750 in 1956 to about 2,000 at the present time. But although the four islands on which the seals breed predominantly are becoming overcrowded, the latter show little inclination to establish new colonies on unoccupied islands. As we have remarked more than once in the preceding books in this series, animals are extraordinarily conservative in their adherence to the traditional breeding and wintering places of their kind, and more often than not no amount of persecution can drive them permanently away from these.

There are as many breeding seals on the Orkney Islands as on the Farnes, but they are scattered in relatively small colonies throughout the Orkneys, while elsewhere in the British Isles they are mainly dispersed in small groups on the most inaccessible islands and beaches, and also in caves—their exclusive breeding places in Iceland and the Faroe Isles—where a colony may consist of its irreducible minimum of bull, cow and pup. These favoured islands and beaches have some special attraction to the breeding cows which is not apparent to the human eye, for at other seasons they haul out on other islands and beaches which appear to be no less suitable for breeding. We have seen that no amount or duration of persecution can drive seals in general from their traditional breeding stations, and if the species is to survive natural hazards, as opposed to the unnatural hazard of persecution, these must therefore be the best places in which to rear their young. In all parts of their range the gray seal pups are born during the stormiest months of the year, though the actual birth months vary from September on south-west and west coasts of Britain—a small number are born late in

March in Wales—to September and October on Rona and from the latter half of October to mid-December on the Farnes. Lockley has pointed out that this variability coincides with those seasons when the fishermen of these localities haul up their boats during the bad-weather season for their annual overhaul, with the result that these are the periods when the seals' breeding places will not be visited; and that moreover this coincidence also holds good for Iceland, the Faroes and Norway, where the seals breed in October. But while this is true of some seal localities in Britain, it can hardly be said to apply to Rona or to the Farnes. Moreover, in the central and northern waters of the Baltic, where perhaps 5,000 or 10,000 gray seals live in a sub-arctic environment and are reported to keep open breathing holes in the ice like the true arctic seals, the breeding season is tied to the seasonal condition of the ice, and the pups are born at annually varying dates from mid-February to April, depending upon the availability of stable drift or fast ice. Similarly, on the American side of the Atlantic, the Canadian population of some 3,000 gray seals haul out to breed in very small colonies from mid-December to early March, and while some pup on snow-covered islands, others do so on the ice. From year to year their breeding season varies, according to the dates on which the pack-ice is blown inshore and becomes cemented into a large field of land-fast ice. These colonies, which are located north and south of, and also in, the Gulf of St. Lawrence, with a small colony as far south as Nantucket, are so inaccessible that one of some 400 seals on Basque Island off Cape Breton Island was only discovered during an aerial survey in 1962, while a previously unrecorded group of cows and pups were spotted on the ice in the Gulf of St. Lawrence in January 1965; and there may be other unknown colonies. The largest known colony, producing 300 pups annually, has been established on the constantly-shifting sand-rib, 25 miles long and 1 mile wide, of Sable Island, that graveyard of wrecked ships 120 miles east of Nova Scotia. Walruses hauled out on Sable Island in the seventeenth century, for though fringed with ceaseless surf it was often ice-free in the spring. In a normal winter any snow that falls is quickly blown away, but many of the seal pups become encrusted with blown sand, and it seems a most unsuitable environment for them.

Gray seals inhabit coastal waters, and probably do not disperse very far from their breeding stations once they are adult. Their food is therefore mainly the rock and bottom fish in depths of up to 250 feet, or perhaps deeper, for one has been hooked on a line at a depth of between 420 and 480 feet—such fish as saithe, pollack, lythe and skate, and other mainly non-commercial fish like wrasse, rays and the lumpsuckers, which Northumbrian fishermen have always declared to be the favourite fish, and which a seal eats in a peculiar way, gripping the lumpsucker by the belly and literally shaking it out of its skin. Whiting and herring are also taken together with conger-eels, cephalopods and some crustaceans, while the more migratory younger seals poach some salmon and sea-trout. The common seals, on the other hand, though also feeding on such fish as whiting, flatties and gobies, and on prawns and shrimps, are mainly shellfish merchants with a special liking for cockles and whelks. There are only about 8,000 common seals in the British Isles, no more than from 2 to 5 per cent perhaps of the combined Atlantic and Pacific populations, which include those on the Canadian arctic islands, Greenland and the Bering Sea.

Gray seals are present throughout the year at accessible breeding stations such as the Farnes, where their numbers are never less than an absolute minimum of 10 per cent of the total population, and where the numbers hauled out on the rocks at one time may indeed exceed 2,000. At storm-bound stations, such as Rona, suitable weather for hauling out is restricted, and though perhaps 10 per cent fish Rona waters throughout the year, it is in July and August, at the approach of the breeding season, that the bulk of the population begins to return to that island. During the off-season they have fished intensively, catching perhaps 10 or 15 pounds of fish a day, and consuming far more than their energy output requires. Thus by the onset of the breeding season they, like the elephant seals, fur seals and sea lions, are exceedingly fat, carrying perhaps 1½ hundredweight of soft fat according to Darling; and it has been estimated that a large bull, approaching 10 feet in length and with a girth of 7½ feet behind the flippers, may weigh as much as 1,000 pounds, and the cows 350 or 400 pounds, compared to the 250 pounds of the 6-feet-long bull common seal, though how does one estimate the weight of a wild seal?

On first returning to Rona the seals haul out to rest on the outer rocks. Although the island is cliff-bound and the immense Atlantic swell makes the sea's edge a dangerous place, a flattish expanse of rock at its northern end is sufficiently broken up for those cows hauling out to retain a hold even when the huge swells are running over it, and not be sucked off in the backwash. Moreover, these rocks afford access by way of gullies to a broad peninsula, and thence up to a moor 300 or 400 yards from the sea and as high as 300 feet above it. Seals, as we have seen, have immense gripping power, and Darling cites one instance of a cow, heavy with pup, 75 feet up a cliff with a 45 degrees gradient, and of another which reared a pup on the brink of a 300-foot column of cliff.

The rocks at the edge, gateway to the breeding grounds on the moor above, are neutral ground, and throughout the breeding season are the headquarters of a reservoir of some 500 unmated bulls; but although incoming cows rest there for from two to five days before moving inland to pup, there is neither courtship with the bulls nor aggression among the latter. It is at the end of August that the first bulls move up on to the moor to take up territories, preferably near shallow dubs of water into which they subside from time to time to wallow and cool off. In so doing, they churn up the muddy bottoms, before flopping out to dry off, and this results in a fascinating ecological relationship, for Darling observed that as soon as the bulls took up territories in September the turnstones (some of which he suspected of breeding on Rona) altered their habits, for the commotion the bulls made in the pools doubtless left many tiny creatures stranded on the edges and the turnstones were not slow to take advantage of this harvest. They also inspected the seals themselves and assiduously turned over any seal excrement they could find. A dub becomes the approximate centre of a bull's territory and also the place in which he is likely to mate with the cows some three weeks later, a fortnight or so after they have pupped. Throughout September the numbers of cows increase, as they land in groups at intervals of a week or so, and the peak of the pupping season is reached in the second week of October. But the bulls do not collect harems, for each of the four or five or twenty cows that have taken up positions in his territory is free to wander where she will without check from him, and belongs temporarily to any bull whose territory she enters, and may even mate with him. Contrast this behaviour with the incessant chivvying and rounding up of cows by bull elephant and fur seals. It seems odd, but it is to his territory that the gray bull is mated rather than to the cows, and one may indeed occasionally wait for weeks for a cow to enter his territory. Yet this same bull, while he drives off any bachelor newly come up from the sea to challenge for ownership of a territory, makes no attempt to encroach upon a neighbouring bull's territory, nor to steal a cow from it. At other British stations the bulls may not even establish territories, mating in the sea or on the rocks. On the Farnes bulls make little attempt to take up territories until the first pups have been born, and at the height of the breeding season I have had three cows alongside my boat courting a bull. Rearing up in the water, they nuzzle his mighty head all at one moment, and then play, or display, about him, barking, and threshing the sea with their gambollings. At small island and beach stations the bulls, like the Californian sea lions, ceaselessly patrol the sea adjacent to their territories, and may only haul out ashore for a few hours at a time.

The attitude of a cow towards its newborn pup varies from one individual to another. Most stay with their pups for the first few hours and perhaps suckle them; others leave theirs soon after birth, returning to suckle them a few hours later. For the first day or two the pups are fed four or five times a day, though the majority of the cows reduce the number of feeds thereafter to two a day, tending to come ashore as the tide flows, and leave again on the ebb. Even blind cows, of which most large colonies contain at least one, follow the same procedure, and a British biologist, K. M. Backhouse, has described how one he watched on an island in the Inner Hebrides, and which, incidentally, raised a foster pup in addition to its own—as gray cows quite often do in crowded colonies—came and went several times a day, despite the fact that her route between the sea and her breeding place was quite complicated:

On entering the water channel through the expanse of offshore platform she swam along the edge of the seaweed as if using her vibrissae to guide her. When she came to a junction she would slow down and test the routes and then having chosen the correct one continue along it. The final approach was across some forty or fifty yards of more or less featureless seaweed-covered platform and then through a narrow cleft between blocks of high rock on to the beach; she managed this last stage with remarkable ease and from the sureness of her approach, it was easy to doubt her blindness even though this was an established fact.

She had no difficulty in finding her pup, any more than a fur seal cow has when, returning periodically to suckle hers, she locates it among the massed thousands of pups and cows in the rookery; for the two home to each other across the beach, first by mutual recognition of voice and then by scent, as does the bull sea lion when investigating the cows, pups and yearlings in his territory. Indeed a fur seal will feed no other pup except her own.

For the first couple of days after birth the thin, 3-feet-long, gray seal pups are very active; but when their skins begin to tighten with fat, they pass much of their time asleep on their backs. Fed on milk that is more than 50 per cent fat and several times as rich as a Jersey cow's milk, the pups put on up to 3¾ pounds in weight a day and quickly become barrels of fat, almost trebling their birth weight of from 25 to 40 pounds in three weeks. Since they are born during the wildest months of the year, their birthplace must be one in which they are protected as far as possible from storm seas; and at first sight a high island such as Rona would appear to be much more suitable than ledges in the deep recesses of caves or small coves or narrow strips of pebble beaches backed by high cliffs, all of which are overrun by high seas pounding on the shore and flooding into the caves at spring-tides; and more suitable than the flattish island rocks of the Farnes over which great seas break in Force 10 gales. Pups born in such situations, at, or only a little above sea-level, are constantly exposed to big seas, for though they can swim in still rock pools when only two or three days old, they are not seaworthy for several days after birth. Such seas wash the pups out of caves and suck them off the beaches, wedging them into crevices from which they cannot subsequently extricate themselves, battering them to death on the rocks, or carrying them away into the breakers. It is true that beach-breeding cows may encourage their pups into the sea as early as the third day after birth, though taking the precaution of swimming between them and the open sea; and on both the Farnes and the Welsh islands cows have been observed taking pups to swim in the sea four or five times a day for periods varying from three-quarters of an hour when four days old to 1½ hours when a week old. But even these precocious pups are not truly seaworthy, and the cows are most heroically and pathetically solicitous for their safety during storms.

A naturalist on the Farne Islands, Eric Simms, when watching a cow giving birth on a November day, noted that she appeared to take no interest in the pup until a breaker swept it into the sea about four minutes after it had been born. It took the cow a couple of minutes to realize what had happened; but then, with a tremendous leap, she dived into the sea and swam to the seaward side of the pup:

> Some twenty minutes later she succeeded, after strenuous efforts, in breasting it against her chest, and between her fore-flippers, into a creek in the rock, over which the sea was breaking. She then lay in the creek and thus allowed the sea

to break on her back, preventing the calf being carried out into the surf again. It was clear that after the twenty minutes period the calf's silky coat was becoming waterlogged, and it was experiencing great difficulty in keeping above the water.

Lying broadside below their pups, the cows break the force of the surf and prevent the pups from being washed away, or attempt again and again to push them back on the beach if they fall into the surf, even lifting them up with their flippers against ledges so that they can endeavour to climb on to them. If storm waves are breaking over a straying pup its mother will seize it by the ankles of its hind flippers and fling it behind her on to the beach, or follow it into the combers if it has been rolled back into deep water and, putting protective flippers around it, draw it back to the beach, or perhaps be carried away with it in her clasp.

Nevertheless, the annual 10 to 25 per cent mortality among pups is no less on Rona than it is on those breeding stations swept by storms, for pups born in large crowded colonies are exposed to dangers not experienced by those born in small colonies. On Rona, where 2,500 pups occupy only one-third of the 300 acres of moor available, many are born and crushed in the traffic lane of bulls coming and going from the sea, for in contrast to the harem seals, individual gray seals do not maintain territories throughout the season, and there is a constant traffic of fresh bulls ascending to the breeding grounds on the moor and of spent ones descending to the sea. Cows breeding in crowded conditions, or subjected to constant human interference as they are on the Farnes, are liable to be aggressive towards each other and also to the pups of other cows straying near them. In these circumstances frightened pups become separated from their mothers, wander about the breeding grounds and, soliciting other cows, are snapped at or actually bitten, or fall into trenches or crevices; while those cows breeding furthest from the sea, and having to pass through the ranks of other aggressive cows, may not be able to suckle their pups often enough during the critical first two days. Starvation is therefore probably the greatest cause of mortality at stations such as Rona or the Farnes.

Paradoxically, the very fact of Rona pups being born in an environment secure from storm hazards, actually prolongs their exposure to crowding hazards. Both gray and common seals are born with coats of silky white hair which are not waterproof. Although in arctic regions some common pups are born on the ice, in Britain most are born on extensive sandbanks, or even in the sea, during the relatively calm months of June and July—at low water on midsummer night say the East Coast fishermen. Since the pups are probably seaworthy within a few hours of birth, the cows can if necessary haul out on to the sands, usually below high-water mark, as the tide falls, give birth and swim off with their pups when the tide rises again; for the latter can suckle as well in the water as out of it—as can the gray pup in exceptional circumstances—and can even do so while being towed along under water, gripping the teat between a special notch in the tongue and the gum between the canines of the upper jaw. However, the majority of the pups are probably born out of the water, and if not disturbed by man or swept away, during their mothers' absence, by strong-running tides or heavy ground-swell on extensive sandbanks such as those of the Wash, are suckled at frequent intervals in their birth-places for several days. Indeed the cow's devotion is constant for the first two or three weeks, and the pup passes much of its time playing in the sea with her, after it has received its first swimming lesson cradled in her flippers or mounted on her back; and when the two rest, they float side by side at the surface, heads slightly awash. But by the time it is three weeks old the pup is already able to sleep under water, with eyes tightly shut, for three minutes at a time, before rising, still asleep, to breathe for three seconds.

Now the common pup, which at birth is only 12 inches shorter and 8 pounds lighter than the gray pup, obviously could not be seaworthy within a few hours of birth if it retained its white natal coat, for this, like the gray pup's, becomes waterlogged during prolonged immersion. But common seals have evolved a remarkable safeguard which permits the pups to moult this first coat *before* birth, so that, as Harrison Matthews has expressed it, "The stage of infantile helplessness has been reduced and, as it were, pushed back in the individual's life so that the most helpless stage is got over before the pup's birth." Moreover, in the insecure environment of the Welsh islands, where in caves and on narrow beaches they are exposed to the hazards of storms and spring-tides, many gray seal pups are also part-moulted at birth, while the remainder begin to shed their birth-coats within seven or fourteen days of birth: but the Rona pups, in their secure environment, do not begin to moult until their fourth week. Although the earlier moult of the Welsh pups may be partly due to their greater activity, in the course of which the fluff is rubbed off, the fact remains that they are fully competent swimmers when little more than a week old, are weaned when from fourteen to eighteen days old—on the day that the cow mates with the bull in some instances—and, like those on the Farnes, are able to go to sea before they are three weeks old; though they may subsequently haul out again on nearby beaches for a few days. The Rona pups, on the other hand, are not weaned until they are three or four weeks old, and do not begin their long and difficult journey to the sea, which may involve negotiating 75-foot sheer cliff drops, for a further week or two. Thus, while the Welsh pups, in their insecure habitat, may have to accelerate their moult in order to cut short the period of unseaworthiness, the slower maturing Rona pups have to pay for their security from the sea by longer exposure to the hazards of crowded colony life. Although some pups on the Farnes tend to form groups of seven or eight after they have been weaned and deserted by their mothers, most gray pups are solitary natured, unlike the common seal pups which join up in play communities with other pups on the strand well above low-water mark and in weedy pools, wherein they make their first independent assays at catching their own food of prawns, shrimps, and small rock fish.

By the time it is weaned and abandoned by its mother the average gray pup weighs no less than 98 pounds, with males about 5 pounds heavier than females; but during the period between weaning and its departure for sea it loses weight at a rate of rather more than 1 pound a day, and no doubt continues to lose weight for some time thereafter, until able to obtain adequate food by its own efforts; and indeed those nine-month-old yearlings that have been weighed have proved to be very little heavier than pups. During its first year a pup wanders further afield than at any other time in its life. Pups marked on Rona have wandered to Iceland, Norway and the Frisian Islands off the Netherlands-German coast in the autumn of their birth; and Welsh pups to the west of Ireland and Brittany and, in one instance, the north coast of Spain, which was reached by a ten-week-old pup in good condition after a voyage of 600 miles by the most direct route, though much further by a wind and current driven course. Much the largest number of pups have been marked on the Farne Islands—some thousands indeed during the past twenty years—and the remarkable proportion of over 10 per cent of those subsequently recovered, mostly alive and well, have crossed the North Sea to, in particular, the coasts of Holland, Germany and Norway; one, however, reached the north-west Swedish coast and a second the Faroe Islands, while a third was washed up in the tangled nets of a Soviet trawler off Bergen. The longest voyage of 600 miles has been that of a 8½-week-old pup which reached the Norwegian coast slightly north of Trondheim, and the most precocious that of one which crossed to the east Frisian Islands, 350 miles from the Farnes, when only seventeen days old, though it could not have completed its first moult at that tender age. The remarkably high proportion of European recoveries does not of course indicate a purposeful trans-North Sea migration. These recoveries are those of pups that, while wandering off the easterly seaboard of Britain, have been blown out to sea by westerly gales and caught up in a branch of the North Atlantic Drift current which, after flowing south down the east coast of Britain, swings eastward and northward to Norway and the Arctic. The random nature of their dispersal is indicated by the movements of four other pups which were marked on the same day as the above-mentioned seventeen-day-old, for while a second also reached the east Frisian Islands, a third was reported from Aberdeen and a fourth from Kent, while a fifth remained in Northumbrian waters. Other examples of this random dispersal are provided by a 6½-week-old pup which, four days after being reported on the Yorkshire coast south of the Farnes, was recovered on the coast of Noord Holland 200 miles distant; while another pup, first sighted north of the Farnes—at the Isle of May where there is a small colony of gray seals—when only just over four weeks old, was reported nine days later from a point off the Norwegian coast 360 miles distant.

None of these trans-North Sea wanderers have ever been subsequently recovered at British breeding stations, and it is possible that they and other older immatures have assisted in building up the stocks of European gray seals. On the Norwegian coast, for example, gray seals had been almost exterminated by hunters, but there has recently been a dramatic increase to a total population of perhaps 3,000, with breeding colonies as far north as the Murman coast at the entrance to the White Sea, and this increase has been attributed to colonization by British seals. However, the mortality rate from such hazards as storms, killer whales, sharks, fishing nets, polluted waters and the man with the gun is so severe during the pups' first year of life, when as high a proportion as 60 per cent may be lost, that this seems unlikely, though it is true that there has been some colonization in British waters, with a marked increase in the numbers of gray seals on Orkney and Shetland during the past thirty years and a new colony founded on St. Kilda within the last twenty years. Moreover, any colonization by young seals must be very slow to take effect, for though some Welsh pups have been reported at the breeding stations in the spring following their birth, the majority of those that survive do not return until four years old, unlike the common seal pups which apparently do not wander far from their birthplace and herd with the adults when the latter haul out to moult in the autumn. Furthermore, the gray cows do not come into breeding condition until from 4 to 6 years old, and the bulls are not sexually mature until perhaps 9 or 10, though they may have another 30 years of life before them. Nor is there any evidence that they wander far from their breeding stations once they are adult. Indeed one bull, marked as a pup on Ramsey Island off the Welsh coast, was recovered fourteen years later actually on the island, and another only 28 miles away after thirteen years.

And so we can leave the gray seals in their sanctuary on North Rona. During the breeding season both cows and bulls have lost a great deal of weight, the cows probably as much as, or more than, one-third. But before they finally depart from the island for seven or eight months fishing, they moult, though at more accessible stations elsewhere in the British Isles the moult is delayed until the New Year or early spring. During late November and early December 2,000 or 3,000 moulting seals are to be seen lying peaceably together on the Rona skerries; but by the middle of December when, during storms, Darling watched the spray breaking the full 300 feet over the western cliffs, only a few late pups and spent bulls remain on the island.

In Conclusion

Throughout this book we have been constantly aware of curious, even bizarre relationships between the most diverse forms of life—crabs and coconut palms, flamingoes and bacteria, marine turtles and jungle jaguars and tigers. We have also examined many examples of intricate ecological relationships—sea-urchins, kelp forests and sea otters, oceanic sea-birds and reptilian tuataras—and of such relationships destroyed by man, as in the case of the Everglades snails and the kites dependent on them. But let us conclude by tracing the ecological links between rabbits, sea-birds and seals.

The dramatic increase in the tonnage of gray seals breeding on the Farne Islands is resulting in a drastic erosion of the thin layer of soil covering these rocky islands, and thereby putting at risk the continued existence of such burrow-nesting sea-birds as puffins, whose numbers are already declining catastrophically on other British islands from causes

Rabbit, terns, and puffin in competition on the Farne Islands

by September the only remaining green vegetation is the sea-campion, which cannot be classed as normal rabbit food. Finally, from late October until the end of the year thousands of seals are churning the herbage floor into a morass, and gales and drenching salt-spray complete the devastation. How, not fully established, though rats, gulls and pollution have certainly contributed to this decline. On two of the Farne Islands (connected at low water) on which puffins nest and which are also the main breeding stations of the seals, there has existed for an unknown number of years a small but unique population of reddish or ginger-coloured rabbits. Happily, a warden on the Farnes, John Cranham, was inspired to undertake a field study of these rabbits, with most enlightening results, which he has summed up in *Animals* magazine. If ever there was an example of the survival of the fittest, it is that of these Farne rabbits, whose numbers Cranham found to fluctuate from as few as eight in the winter to ten times as many in the sumer. From the days of the first litters of young, early in April, the rabbits are engaged in a ceaseless struggle to survive, for the birth of the young coincides with the return of the puffins and severe competition for burrows. In Cranham's experience the puffins are successful more often than not, with the result that the majority of the rabbits are allowed to occupy only those burrows that the puffins do not require. The question is, then, will the seals' erosion of the islands' crust benefit the rabbits by rendering it unsuitable for puffins, or also make it untenable for the rabbits?

The return of the puffins is, however, only the prelude to the rabbits' trials, for no sooner have the young been born than the does, now obliged to spend much of their time feeding above ground, are exposed to continuous attacks by the sharp-beaked terns on whose nesting grounds they encroach while, later, herring and lesser black-backed gulls take their toll of the young rabbits when these are old enough to venture out of their burrows. Furthermore, by July the luxuriant spring vegetation which includes the rabbits' favourite chickweed, is withering on the thin porous soil:

then, are rabbits to avoid starvation during the winter and early spring? A very limited supply of food can no doubt be obtained from plant roots, but Cranham observed that during spells of fair weather in the winter the rabbits were to be found, day and night, feeding on seaweed, whether growing on the rocks or washed up by storms. Seaweed, of course, has some nutritional value. The channelled wrack's mat of short, multi-branched fronds, which grows in a narrow band near and just below high water, serves, in particular, as a subsistence fare for ponies, cattle and sheep along many stretches of coast in the Shetland Isles, the Hebrides and the Scottish Highlands.

In addition to seaweed, Cranham's rabbits grazed the lichens on the rocks and gnawed at driftwood, providing that this had not been tarred or creosoted. The Farne Islands' population has therefore been able to survive against all odds primarily by beachcombing during the winter. However, it only just survives for, at the end of the winter, there may be fewer than ten individuals remaining. Moreover, Cranham found that in two study seasons the average litter comprised only four young, and the average doe had produced only nine young with the completion of her last litter early in July.

From an ecological angle even more interesting than the rabbits' survival is Cranham's argument that they present a perfect example of natural population control, and that any attempt to improve their state would be disastrous not only for them but also for other inhabitants of the Farnes. He argues that any artificial enrichment of the vegetation would inevitably result in the numbers of rabbits increasing during the summer to such an extent that the bare minimum of food available during the winter would be over-exploited, with the result that all might starve. If, on the other hand, the rabbits were removed from the islands, the herbage complex would be radically changed. The Inner Farne, the largest of the islands, bears witness to this. In the autumn of 1968 the feral gray and black-and-white descendants of stock introduced by generations of lighthouse keepers were exterminated, and during the next two years Cranham observed that the various flowering plants, and in particular thrift, grew more profusely than previously, while clover and dandelions reappeared; but in the third year docks, nettles and such coarse grasses as Yorkshire fog rampaged, crowding out many of the weaker but more valuable plants. As a result of this change in the vegetational pattern, and also the density of new growth, the three species of terns—common, arctic and roseate—which habitually nested on the rabbit-controlled pastures, were unable to continue doing so; and their colonies were further reduced by the heavy mortality among chicks, unable to escape out of the jungle of thick wet grass during rainy periods, and dying from exposure. In place of rabbits, chemical sprays and motor mowers are now required to control the herbage in the tern colonies!

But consider, by contrast, the situation on St. Kilda. Since the evacuation of the St. Kildans more than forty years ago their sheep stock has bred and grazed uncontrolled, with the result that the herbage on the three main islands now consists of a fine short greensward. Consequently such burrow-nesting sea-birds as puffins, petrels, and shearwaters are restricted to steep cliff slopes and lesser islets, because adults returning to their burrows, and young leaving them, have been deprived of cover in which to conceal themselves from increasing numbers of large gulls.

Here then, in a nutshell, is a summary of the infinitely complex inter-relationships embraced by the term "ecology".

Selected Bibliography

Ali, Salim. "More About the Flamingoes in Cutch." *Journal of the Bombay Natural History Society* 45 (1945): 586–92.

Allen, R. P. *The Flamingoes*. National Audubon Society research rept. 5, 1956.

Attenborough, David. *Zoo Quest for a Dragon*. London: Lutterworth, 1957.

———. *Zoo Quest to Madagascar*. London: Lutterworth, 1961.

Auffenberg, Walter. "A Day with Number 19." *Animal Kingdom*, December 1970: 19–23.

———. "Komodo Dragons." *Natural History* 81 (1972): 52–59.

Backhouse, K. M. *Seals*. New York: Golden Press; London: Barker, 1969.

Bailey, A. M. "The Hawaiian Monk Seal." *Museum Pictorial* 7 (1952): 1–32.

Bannerman, David A., and Lodge, George E. *The Birds of the British Isles*, vols. 6 and 9. Edinburgh: Oliver & Boyd, 1957 and 1961.

Barabash-Nikiporov, I. I. *The Sea Otter*. Trans. by Israel Program for Scientific Translations. Washington, D.C.: National Science Foundation, 1962.

Bárdarson, Hjálmar R. *Ice and Fire: Contrasts of Icelandic Nature.* Trans. by Sölviz Eysteinsson. Reykjavik: Hjálmar R. Bárdarson, 1971.

Bates, Marston. *Animal Worlds.* New York: Random House; London: Nelson, 1963.

———, and Abbott, Donald P. *Ifaluk: Portrait of a Coral Island.* London: Museum Press, 1959.

Beamish, Tony. *Aldabra Alone.* London: Allen & Unwin, 1970.

Beck, Rollo H. Expedition of the California Academy of Sciences 1905–1906. *Proceedings of the California Academy of Science* 1–11 (1907–18).

Beebe, William. *Arcturus Adventure: An Account of the New York Zoological Society's First Oceanographic Expedition.* New York and London: Putnam, 1926.

———. *Galápagos: World's End.* New York and London: Putnam, 1924.

Bellairs, Angus, and Carrington, Richard. *The World of Reptiles.* London: Chatto & Windus, 1966.

Bernstein, Joseph. "Tsunamis." *Scientific American* 191 (1954): 60–64.

Bourne, Arthur. "Surtsey Comes to Life." *Animals* 11 (1968): 179.

———. "Surtsey: The First Life Moves In." *Animals* 8 (1966): 492–95.

———. "Surtsey: The Island Evolves." *Animals* 8 (1966): 436–39.

———. "Surtsey: The Volcano That Became an Island." *Animals* 8 (1966): 314–15.

Bowman, Robert I. *The Galápagos.* Berkeley and Los Angeles: University of California Press, 1966.

Bristowe, William S. *A Book of Islands.* London: Bell, 1969.

Brown, Leslie. *Africa: A Natural History.* New York: Random House; London: Hamish Hamilton, 1965.

———. *The Mystery of the Flamingoes.* London: Country Life, 1959.

Burden, Douglas W. *Dragon Lizards of Komodo.* New York and London: Putnam, 1927.

Carlquist, Sherwin. *Island Life: A Natural History of the Islands of the World.* New York: Natural History, 1965.

Carpenter, Charles C. "Notes on the Behaviour and Ecology of the Galápagos Tortoise on Santa Cruz Island." *Proceedings of the Oklahoma Academy of Science* 46 (1966): 28–32.

Carr, Archie Fairly. *So Excellent a Fishe: A Natural History of Sea*

Turtles. New York: Natural History Press, 1967. (Published in Great Britain as *The Turtle: A Natural History of Sea Turtles.* London: Cassell, 1968.)

———, and the Editors of *Life. The Reptiles.* New York: Time, Inc., 1964.

Chapman, Abel, and Buck, Walter J. *Unexplored Spain.* London: Arnold, 1910.

———. *Wild Spain.* London: Gurney & Jackson, 1893.

Chapman, Frank M. *Camps and Cruises of an Ornithologist.* London: Hodder & Stoughton, 1908.

———. "A Contribution to the Life History of the American Flamingo." *Bulletin of American Museum of Natural History* 21 (1905): 53–77.

Chubb, Lawrence John. *Geology of the Galápagos, Cocos and Easter Islands.* Honolulu: Bishop Museum, 1933.

Colinvaux, Paul A. "Eruption on Narborough." *Animals* 11 (1968): 297–301.

Couffer, Jack. *Song of Wild Laughter.* New York: Simon & Schuster, 1955; London: Constable, 1963.

Cousteau, Jacques-Yves, and Diolé, Philippe. *Life and Death in a Coral Sea.* Trans. by Jack F. Bernard. Garden City, N.Y.: Doubleday; London: Cassell, 1971.

Cowley, Ambrose. *Voyage Round the World.* n.p., 1699.

Cranham, John. "Farne Island Rabbits." *Animals* 14 (1972): 244–45.

Cristopherson, Erling. *Tristan da Cunha: The Lonely Isle.* Trans. by R. L. Benham. London: Cassell, 1940.

Darling, Frank Fraser. *Natural History in the Highlands and Islands.* London: Collins, 1947.

———. *A Naturalist on Rona: Essays of a Biologist in Isolation.* Oxford: Oxford University Press, 1939.

Darling, Lois, and Darling, Louis. *Coral Reefs.* Cleveland: World, 1963; London: Methuen, 1965.

Darwin, Charles. *Journal of Researches into the Geology and Natural History of the Various Countries Visited During the Voyage of H.M.S. "Beagle" Round the World.* London: Dent, 1906.

Davies, J. L. "Observations on the Grey Seal at Ramsey Island." *Proceedings of the Zoological Society of London* 119 (1949): 673–92.

Dröscher, Vitus B. *The Magic of Senses: New Discoveries in Animal*

Perception. Trans. by Ursula Lahrburger and Oliver Coburn. New York: Dutton; London: W. H. Allen, 1969.

Eibl-Eibesfeldt, Irenaus. *Galápagos: The Noah's Ark of the Pacific*. Trans. by Alan Houghton. New York: Doubleday, 1961; London: MacGibbon & Kee, 1960.

———. *Land of a Thousand Atolls: A Study of Marine Life in the Maldive and Nicobar Islands*. Trans. by Gwynne Vevers. New York: International Publications, 1966; London: MacGibbon & Kee, 1965.

Fisher, Edna M. "Sea Otters." *Journal of Mammalogy* 20 (1939): 21–36.

Fisher, James. *Rockall*. London: Geoffrey Bles, 1956.

Gallet, Étienne L. *The Flamingoes of the Camargue*. Oxford: Blackwell, 1950.

Garman, S. *The Galápagos Tortoises*. Cambridge: Cambridge University Press, 1917.

Gaymer, Roger. "Aldabra's Giant Tortoises." *Animals* 10 (1967): 192–93.

———. "The Indian Ocean Giant Tortoise on Aldabra." *Journal of the Zoological Society of London* 154 (1968): 341–63.

Gillsäter, Sven. *From Island to Island: Oases of the Animal World in the Western Hemisphere*. Trans. by Joan Tate. London: Allen & Unwin, 1968.

———. *We Ended in Bali*. Trans. by F. H. Lyon. London: Allen & Unwin, 1961.

Guggisberg, C. A. W. *Crocodiles: Their Natural History, Folklore and Conservation*. Newton Abbot, Devon: David & Charles, 1972.

Guppy, H. B. *The Solomon Islands and Their Natives*. London: Swann Sonnenschein & Lowry, 1887.

Hachisuka, Masauji. *The Dodo and Kindred Birds: or, The Extinct Birds of the Mascarene Islands*. London: Witherby, 1953.

Hagen, Victor Wolfgang von. *Ecuador and the Galápagos Islands*. Norman: University of Oklahoma Press, 1949.

———. "The Flamingoes of the Galápagos Islands." *Natural History* 39 (1937): 137–39.

Hall, K. R. L., and Schaller, G. B. "Tool-Using Behaviour of the Californian Sea Otter." *Journal of Mammalogy* 45 (1964): 287–98.

Hancock, James A. "Everglades Crisis." *Animals* 10 (1968): 566–67.

———. "Men Versus Alligators." *Animals* 9 (1966): 489–93.

Hanna, G. Dallas. "Rare Mammals of the Pribilof Islands, Alaska." *Journal of Mammalogy* 4 (1923): 209–15.

Harris, C. J. *Otters: A Study of the Recent Lutrinae*. London: Weidenfeld & Nicolson, 1968.

Hass, Hans. *Expedition into the Unknown: A Report on the Expedition of the Research Ship "Xarifa" to the Maldive and Nicobar Islands*. Trans. by Gwynne Vevers. London: Hutchinson, 1965.

Hewer, H. R. "The Grey Seal on North Rona." *New Scientist* 6 (1959): 158.

———. "A Hebridean Breeding Colony of Grey Seals." *Proceedings of the Zoological Society of London* 128 (1957): 23–66.

Heyerdahl, Thor. *Sea Routes to Polynesia*. Chicago: Rand McNally; London: Allen & Unwin, 1968.

Hickling, Grace. *Grey Seals and the Farne Islands*. London: Routledge & Kegan Paul, 1962.

Hobson, Edmund S. "Forests Beneath the Sea." *Animals* 7 (1965): 507–11.

———. "Observations on Diving in the Galápagos Marine Iguana." *Copeia* 2 (1965): 249–50.

Keast, Allen. *Australia and the Pacific Islands: A Natural History*. New York: Random House; London: Hamish Hamilton, 1966.

Kenyon, Karl W. "Return of the Sea Otter." *National Geographic* 140, 4 (1971).

———. "The Sea Otter." *Report of the Smithsonian Institution 1958*: 399–407.

King, F. Wayne. "Ora: Giants of Komodo." *Animal Kingdom*, (August 1968): 2–9.

Klingel, Gilbert. *Inagua: Which Is the Name of a Very Lonely and Nearly Forgotten Island*. New York: Dodd, Mead, 1940; London: Robert Hale, 1942.

Lack, David. *Darwin's Finches*. New York: Macmillan; Cambridge: Cambridge University Press, 1947.

Linblad, Jan. *Journey to Red Birds*. Trans. by Gwynne Vevers. New York: Hill & Wang; London: Collins, 1969.

Lockley, Ronald M. *Grey Seal, Common Seal: An Account of the Life Histories of British Seals*. New York: October House; London: André Deutsch, 1966.

———. *Seals and the Curragh.* New York: Devin-Adair, 1955. London: Dent, 1954.

Loveridge, Arthur. *Reptiles of the Pacific World.* New York: Macmillan, 1945.

Lucas, Frederick A. "Historic Tortoises and Other Aged Animals." *Natural History* 22 (1922): 301–5.

McCann, Charles. "The Flamingo." *Journal of the Bombay Natural History Society* 4 (1939): 12–38.

MacNae, William. "A General Account of the Fauna and Flora of Mangrove Swamps and Forests in the Indo-West Pacific Region." *Advances in Marine Biology* 6 (1968): London and New York: Academic Press.

———, and Kalk, Margaret. *Natural History of Inhaca Island.* Johannesburg: Witwatersrand.

Matthews, L. Harrison. *British Mammals.* London: Collins, 1952.

Melville, Herman. "The Encantadas: or, The Enchanted Isles." In *The Piazza Tales,* 1856. London: Constable, 1929.

Milne, Lorus, and Milne, Margery. *The Senses of Animals and Men.* New York: Atheneum; London: André Deutsch, 1963.

Morrison, Tony. "Three Flamingoes of the High Andes." *Animals* 11 (1968): 305–9.

Mountfort, Guy. *Portrait of a Wilderness: The Story of the Coto Doñana Expeditions.* London: Hutchinson, 1958.

———. *The Vanishing Jungle: Two Wildlife Expeditions to Pakistan.* London: Collins, 1969.

Nelson, Bryan. *Galápagos: Island of Birds.* London: Longmans, 1968.

Newman, K. B. "Vanishing Birds of the Seychelles." *Animals* 5 (1965): 475–81.

Nicholson, E. M.; Ferguson-Lees, I. James; and Hollom, Philip A. D. "The Camargue and the Coto Doñana." *British Birds* 50 (1957): 497–519.

Norton-Griffiths, M. "The Dexterous Oystercatcher." *Animals* 8 (1965): 92–95.

Ommanney, Francis D. *The Shoals of Capricorn.* New York: Harcourt, Brace; London: Longmans, Green, 1952.

Orr, Robert T. *Animals in Migration.* New York: Macmillan, 1970.

———. "The Galápagos Sea Lion." *Journal of Mammalogy* 48 (1967): 62–69.

Parsons, James Jerome. *The Green Turtle and Man.* Gainesville: University of Florida Press, 1962.

Percy, William. *Three Studies in Bird Character: Bitterns, Herons, and Water Rails.* London: Country Life, 1951.

Perry, Richard. *At the Turn of the Tide*, rev. ed. New York: Taplinger, 1972; London: Croom Helm, 1973.

———. *Lundy: Isle of Puffins.* London: Lindsay Drummond, 1940.

———. *A Naturalist on Lindisfarne.* London: Lindsay Drummond, 1946.

———. *The Polar Worlds.* New York: Taplinger; Newton Abbot, Devon: David & Charles, 1973.

———. *The Unknown Ocean.* New York: Taplinger; Newton Abbot, Devon: David & Charles, 1972.

———. *The World of the Jaguar.* New York: Taplinger; Newton Abbot, Devon: David & Charles, 1970.

———. *The World of the Tiger.* New York: Atheneum, 1965; London: Cassell, 1964.

———. *The World of the Walrus.* New York: Taplinger; London: Cassell, 1967.

Peterson, Roger Tory, and Fisher, James. *Wild America.* Boston: Houghton Mifflin, 1955; London: Collins, 1956.

Pizzey, Graham. "The Great Barrier Reef." *Animals* 8 (1966): 423–33.

———. "A Visit to Dunk Island." *Animals* 8 (1966): 555–59.

Prebble, Edward A. "Mammals of the Pribilof Islands." *North American Fauna* 46 (1923): 102–20.

Ratcliffe, Francis. *Flying Fox and Drifting Sand: The Adventures of a Biologist in Australia.* London: Angus & Robertson, 1948.

Reader's Digest Association. *The Living World of Animals.* New York and London: Reader's Digest Association, 1970.

Schroeder, Robert E. *Something Rich and Strange.* New York: Harper & Row, 1965; London: Allen & Unwin, 1967.

Seshadri, Balakrishna. *The Twilight of India's Wildlife.* New York: Fernhill House; London: Baker, 1969.

Sharell, Richard. *The Tuatara, Lizards and Frogs of New Zealand.* London: Collins, 1966.

Simms, Eric. In *Transactions of the Natural History Society of Northumberland, Durham & Newcastle upon Tyne* 11, 9 (1956).

Slater, Pat. "The Lacepedes: Bird Islands of North-West Australia." *Animals* 5 (1964): 271–74.

Snow, David W. "The Giant Tortoises of the Galápagos Islands: Their Present Status and Future Chances." *Oryx* 7, no. 6 (1964): 277–90.

Stebbins, Robert, and Kalk, Margaret. "Observations on the Natural History of the Mud-Skipper, *Periophthalmus sobrinus*." *Copeia* 1 (1961): 18–27.

Stoddart, D. R., and Wright, C. A. "Ecology of Aldabra Atoll." *Nature* 213 (1967): 1174.

Straughan, Robert P. L. *Exploring the Reef*. New York: A. S. Barnes; London: Kaye & Ward, 1968.

Stubbs, Tom. "Marine Turtles and the Threat of Extinction." *Animals* 13 (1971): 715–18.

Travis, William. *Beyond the Reefs*. London: Allen & Unwin, 1959.

Turner, D. A. "Flamingoes at Lake Magadi." *Animals* 5 (1964): 340–42.

Valverde, J. Antonio. "An Ecological Sketch of the Coto Doñana." *British Birds* 51 (1958): 1–23.

Venables, U. N., and Venables, L. S. V. "Observations on a Breeding Colony of the Seal Phoca vitulina in Shetland." *Proceedings of the Zoological Society of London* 125 (1955): 531–32.

Weber, Karl, and Hoffmann, Lucas. *Camargue: The Soul of a Wilderness*. Trans. by Ewald Osers. London: Harrap, 1970.

Wetmore, Alexander. In *Life Histories of North American Shore Birds*. U.S. National Museum bull. 146, 2, 1929.

Wiens, Harold J. *Atoll Environment and Ecology*. New Haven: Yale University Press, 1965.

Williams, Hill. "Wildlife in the Thermonuclear Age." *Animals* 9 (1967): 552–55.

Yeates, George K. *Bird Life in Two Deltas*. London: Faber & Faber, 1947.

———. *Flamingo City*. London: Country Life, 1950.

Yonge, C. M. *The Sea Shore*. London: Collins, 1949.

Zahl, Paul A. *Coro-Coro: The World of the Scarlet Ibis*. Indianapolis: Bobbs-Merrill, 1954; London: Hammond, Hammond, 1955.

———. *Flamingo Hunt*. Indianapolis: Bobbs-Merrill, 1952; London: Hammond, Hammond, 1953.

Index